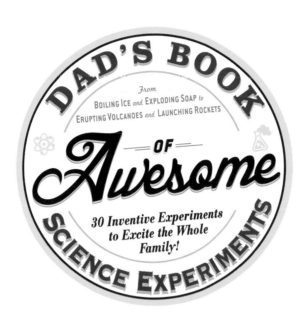

DAD'S BOOK

From
BOILING ICE *and* EXPLODING SOAP *to*
ERUPTING VOLCANOES *and* LAUNCHING ROCKETS

— OF —

Awesome

*30 Inventive Experiments
to Excite the Whole
Family!*

SCIENCE EXPERIMENTS

For Emmeline, my sidekick and inspiration,
and for Dana, who puts up with us.

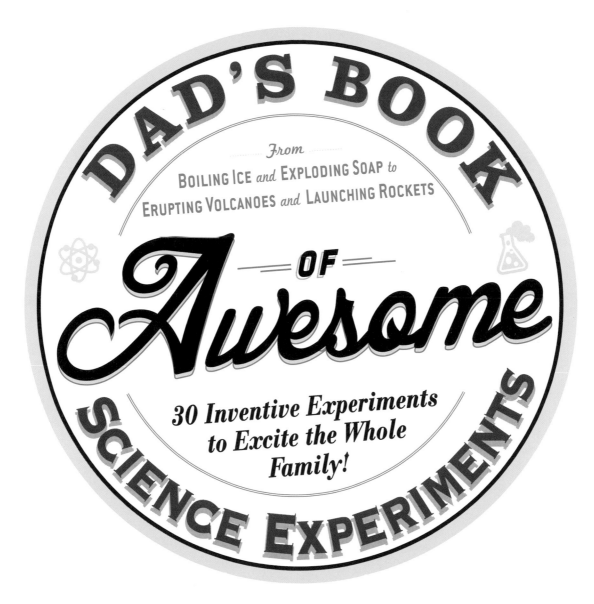

DAD'S BOOK

From

BOILING ICE and EXPLODING SOAP to
ERUPTING VOLCANOES and LAUNCHING ROCKETS

OF

Awesome

30 Inventive Experiments
to Excite the Whole
Family!

SCIENCE EXPERIMENTS

Mike Adamick

Author of *Dad's Book of Awesome Projects*

Aadamsmedia
Avon, Massachusetts

Published by
Adams Media, a division of F+W Media, Inc.
57 Littlefield Street, Avon, MA 02322. U.S.A.
www.adamsmedia.com

ISBN 10: 1-4405-7077-9
ISBN 13: 978-1-4405-7077-3
eISBN 10: 1-4405-7078-7
eISBN 13: 978-1-4405-7078-0

Printed in the United States of America.

10   9   8   7   6   5   4   3   2   1

Library of Congress Cataloging-in-Publication Data

Adamick, Mike, author.
   Dad's book of awesome science experiments / Mike Adamick.
      pages cm
   Includes bibliographical references and index.
   ISBN 978-1-4405-7077-3 (pb : alk. paper) -- ISBN 1-4405-7077-9 (pb : alk. paper) -- ISBN 978-1-4405-7078-0
(ebook)
-- ISBN 1-4405-7078-7 (ebook)
   1. Science--Experiments--Juvenile literature. 2. Science projects--Juvenile literature. I. Title.
   Q164.A327 2014
   507.8--dc23

                                 2013044613

Cover photos by Mike Adamick, illustrations © 123rf.com/ivaleksa.
Photos by Mike Adamick, duct tape images © Hannuviitanen /Bigstockphoto.com, photos on pages 25 (top), 64,
and all other background images and illustrations © 123rf.com/igorr/Oscar Ruelas/ivaleksa/modusuper5.

This book is available at quantity discounts for bulk purchases.
For information, please call 1-800-289-0963.

CONTENTS

Introduction • 9

1 Have Fun, Fail, Learn, and Try Again • 12

2 Chemistry • 20

Soap Clouds • 23
Floating Grape • 27
Penny Shiners • 31
Boiling Ice • 37
Rock Candy Crystals • 41
Rainbow Water Stacks • 47

3 Biology • 54

Colored Leaves • 57
Light Fright • 61
Falling Leaves • 67
Hole-y Walls • 71
Animal Camouflage • 77
Banana Balloon • 81

4 Physics • 86

Straw Balloon Rocket Blasters • 89
Floating Water • 93
Fast Corners • 99
Balloon Toss • 103
Mentos and Coke Rocket • 107
Magnetic Fields • 113
Colors of Light • 119
Egg in a Bottle • 123
Balancing Act • 127

5 Planet Earth • 132

Volcano Time! • 135
Acid Rain • 139
Land Warmer • 143
The Space of Air • 149

6 The Human Body • 154

Hot and Cold • 157
Blind Balance • 161
Birdcage • 167
Marshmallow Pulse Keeper • 171
Fingerprint Monsters • 175

Standard U.S./Metric
Measurement Conversions • 180

Index • 183

Acknowledgments

You never know when a fascination with science will strike. An experiment that opens eyes. Or a teacher who opens minds. I'd like to thank my ninth-grade biology teacher, Ed Mandin, for making science cool for me and generations of kids, and for kickstarting what has become a lifelong fascination with discovery.

Thanks to The Animal Company in San Francisco for providing avian helpers and thanks to Bradley and Matthew Parkhouse for being such great little scientists! Same goes to Samantha Leong! Thanks also to Mady Kim, an amazing bubble maker. Special thanks go out as well to Diya and Esha Mittal for helping to make the photos so awesome, despite being squished in the process.

I'd also like to thank my daughter, Emmeline, for teaching me over and over again that it's okay to mess up and make mistakes, as long as you learn from them and keep going. She is the best lab partner in the world. My wife, Dana Kromm, deserves all the credit for this.

I'd also like to acknowledge all the people who put so much of their DNA into a book behind the scenes: my agent, Mollie Glick, and my fabulous editors, Brendan O'Neill and Katie Corcoran Lytle.

And as always, a big thank you to my beloved community of friends at Cry It Out! You make the Internet fun.

INTRODUCTION

My daughter and I like to work on big, daring, and slightly dangerous projects every time she has a school break. For the past several months, Emmeline and I have been trying to land something on the moon. I remember hearing her plans and thinking, "That sounds crazy" and "Let's do it!" Eat your heart out, NASA . . . or Red Bull . . . or whoever is in charge of space travel nowadays.

Now, did I actually believe we'd land something on the moon? I wasn't hopeful. And yet, we began anyway. Who was I to crush her well-thought-out plan? Who knows, maybe she *could* figure out a way to land on the moon. I'll be darned if I'll be the one to get in the way of all the wonderful science—the engineering, the math, the physics, the space travel.

Kids should feel empowered to take the lead, learn a lot along the way about the world around them, and have fun. This is the idea behind *Dad's Book of Awesome Science Experiments*.

Throughout the book, you'll find thirty easy, interactive, and fun science experiments that can be done by even the youngest of children—with a little help from you, of course. After all, what kid hasn't wanted to make an exploding volcano (Volcano Time! in Chapter 5), fire a balloon rocket (Straw Balloon Rocket Blasters in Chapter 4), cover his hands in ink and examine his own fingerprints (Fingerprint Monsters in Chapter 6), or make her own rock candy (Rock Candy Crystals in

Chapter 2)? And you won't learn just how to do the experiments; you'll also learn all you need to know about how and why these experiments work. Before you know it you'll be helping your lab partner learn about everything from nucleation, to inertia, to thermodynamics, to so much more.

Hopefully, the science experiments found throughout this book will help you inspire your kids to take their love of science to the next level. The starter experiments are there and the chapters are rich with information to explain what's happening. But if your little lab partner wants to take a project in a different direction or learn a little bit more about it, by all means, go for it. Who knows where her curiosity will take her or what world-changing discoveries she might one day make?

The experiments in this book will help you and your little lab partner explore everything from physics to biology, chemistry to planet Earth—and you can do almost all of them with materials and supplies you probably already have around the house. Your kitchen, your yard, your neighborhood—you may not realize it, but they provide some of the best science labs available, and you should take full advantage of all they have to offer. But as you decide which experiments you want to take on first, keep in mind that some experiments require a stove, the microwave, or liquids that can be dangerous if swallowed. I didn't include safety precautions in most chapters, because you know what's safe or doable for your own family. If you think you need safety glasses (you probably won't) or gloves, by all means, put them on. If you need to create a zone of safety for boiling experiments, please do. But for the most part, these experiments are simple and safe enough for just about any kid to do on his or her own.

Throughout the experimentation process, let your kids do the work. There are hints and reminders throughout each experiment to urge involvement by your lab partner, but keep this in mind as you're working together: When in doubt, let your little lab partner get her hands dirty or wet. That's what soap is for. And, frankly, that is what this book is all about. I have no doubt that *you* can mix vinegar and baking soda to create a foam eruption, but let your miniature scientist experience that for herself. Let her wield the tools or the chemicals and take the lead. Let her discover the joys of discovery, as you help out from the sidelines, reveling in the joy of having fun and learning together. Too many toys and activities are bubble-wrapped in a shroud of safety. But these easy science experiments should allow your junior mad scientist to roll up his or her sleeves and get messy. And who knows? Maybe you'll soon find yourself in the backyard, staring up at the sky, and wondering just how to reach the moon . . .

HAVE FUN, FAIL, LEARN, AND TRY AGAIN

1

My daughter and I went for ice cream on a hot summer day and happened upon a place that makes ice cream by the batch. While you wait. Using liquid nitrogen. Emmeline placed her order, and we watched as the workers poured cream into an industrial mixer and then flipped a switch. A swirling white cloud suddenly materialized, enveloping the mixer. It was as if the Wizard of Oz had found a new day job. Emme was enthralled.

It was, indeed, quite the show. I'd never seen anything like it. For a long, long time, we stared in awe, watching the cold, cloudy wisps of nitrogen-infused air swirl and dive. When we sat down on a bench with our cones, I braced for what was to come.

"So, daddy?" Emme began, "How does it . . . *work*? What's liquid nitrogen? Can *we* get a liquid nitrogen mixer?"

Anyone who has ever spent time with a kid knows these questions well—or the countless, seemingly endless questions just like them.

"Why is the sky blue?"

"Why does my heart beat?"

"How do volcanoes work?"

"So wait, birds and bees do . . . *what*?"

The curiosity of a child is an astounding, amazing thing to witness. To a child, the world is a fishbowl of experiences to be discovered, probed, wondered at, and explored. There's a raw, wide-eyed fascination for absolutely everything—a day-to-day awe that somehow, sadly, gets beaten out of us as we grow older.

Some of my absolute favorite times with my daughter have come from casual walks around the neighborhood and the impromptu discussions about her limitless discoveries. It thrills me to watch her face as new revelations dawn, as new discoveries are made. It reminds me to pay attention, to probe, to sometimes stop and simply stare.

That's the direction and guiding principle you'll find in this book: to help turn the everyday into a moment of discovery, to help probe those random, joyful questions, and to help cultivate that awe-inspiring sense of curiosity.

Now, I'm a big believer that you don't need to go out and buy stuff to perform wonderful science experiments or satisfy that endless sense of curiosity. You don't need that fancy science kit or that expensive toy. To be sure, as your kids get older and want to continue their journeys of discovery, there are definitely science supplies to collect along the way. But on this, the start of that journey of discovery, you probably already have everything you need to do some amazing, eye-popping things in your very own home. For instance, if you have some batteries, a screwdriver, and some wire floating around a junk drawer somewhere, you have everything you need to create your very own electromagnet. If you have some fruit and a few party balloons, you can explore the fascinating world of decomposition. Some mints and soda can make for a supersweet backyard rocket and a discussion about propulsion. A coffee filter, some nail polish remover, and a few spinach leaves left over from dinner give you everything you need to discover the wonders of photosynthesis and how the chemical chlorophyll is responsible for the seemingly magical change of tree colors each year.

Each of the thirty experiments found throughout will help you craft easy-to-do science projects with stuff you already have so that you can help explain the world around you or build upon the natural, unbridled curiosity you are witness to every day. But here's the catch. You're going to have fun. You're going to learn all about STEM, or the Science, Technology, Engineering, and Math fields. You're going to get the Scientific Method down pat. And you're going to fail.

The Importance of Failure

Not all of these experiments are *super* easy. Not all of them are quick. Some are made for young children and others are meant for older kids. You're going to get some right, but if you're anything like me, you're also going to fail. A lot.

And that is *great*.

Learning how to move forward after failure is incredibly important for kids—sometimes more important than getting it right. It's so important that in some of the following chapters, we are upfront with them about experiments we tried and failed at before getting right. Here's why: Not everything is easy. Some things require hard work, determination, perseverance, and good ol' fashioned grit. They require the ability to fail, study what happened, and try again. And then again. And again.

We live in a culture rich with information, rife with products, and littered with gadgets that, during our own childhood, we would have seen as pure magic. Need to quickly know why the sky is blue? Slip a hand into your pocket, pull out a phone, and voilà, the information is at your fingertips. We also live in a culture, for better or worse, that often presents images of perfection—whether it's bodies in magazines or craft projects on the Internet. What we don't see is the work behind the scenes: the airbrushing or Photoshopping or the countless multitudes of failures that went into that phone or that craft project. Everything seems so *easy*.

But science, ahh science, it challenges you. It raises questions and begs you to explore them. Science lore abounds with tales of scientists who failed and failed again before making world-bending discoveries: Marie Curie and her radioactivity, Thomas Edison and his light bulb, that dude who invented Post-its. (What? I use them all the time.)

Look, not every kid out there is going to make world-bending discoveries. But early explorations with the experiments found in the following chapters can lay a powerful groundwork for a sense of perseverance, for that roll-up-the-sleeves, buckle-down ability to strive in the face of failure. It can cultivate that natural curiosity that makes life so fun.

It may seem like some sappy Hallmark card to say it's not whether you fail, it's whether you try again afterward—but it's true. And doing hands-on experiments like these can help foster that sense of perseverance, that ability to dig deep and keep going. This type of grit is an important life skill that can get lost in an age of ease and convenience.

You're Going to Have a Blast

Now, while I definitely want to set the expectation that sometimes you'll have to put an experiment on repeat until you get it right, I also want to establish the expectation that you and your lab partner are both going to have a great time. That's the key. That's what it's all about.

You're going to dig up worms in the backyard and dip your hands in freezing-cold bowls of water. You're going to giggle like crazy in the living room as you experiment with balance, and you're going to scream with delight and, possibly, terror as you discover the wonders of inertia on tight turns in the car.

When doing experiments with your kid, it's important to build in a little extra time   more than you think is necessary—because if your little scientist is anything like mine, she's going to want to take the project to another level when you're finished. And that's great. Roll with it. Let her take the lead. And then stand back and watch the gears clink and turn. I promise, there will be moments when you step back and watch as your little scientist is just absorbed, transfixed. Not *every* experiment is going to be like that. But sometimes, magic happens and you'll find yourselves giddy with excitement, pulled closer by the draw of science. Go with it.

Science Is for Everyone

If you have a child in school, you're probably familiar with the STEM movement to raise awareness and student achievement in Science, Technology, Engineering, and Math. The National Math and Science Initiative reports that 54 percent of high school graduates aren't ready for college math, and that 70 percent aren't ready for college science. In addition, while women outnumber men in obtaining college degrees, women only earn 20 percent of engineering degrees. As a parent, you may have heard or seen stories about young girls who were interested in science early on but over the years developed a sense that science was really "just for boys." By doing these experiments with your kids—both girls *and* boys—you lay the groundwork for a broad-based education that will do our society good. You're at the start of the journey. Maybe these experiments can help your child find that spark that propels him or her to a life of joy, discovery, and scientific exploration. Who knows, maybe spending time making Soap Clouds, blowing up a balloon with a banana, or forcing an egg into a bottle will inspire your child to grow up and become an engineer, a chemist, a mathematician. And even if your child decides to one day go into the arts or sports or janitorial services, you'll have given her the kind of expansive, classical education that will help her explore her interests and ensure that all fields feel open to her. What parent wouldn't want that?

The Scientific Method

And speaking of good starts . . .

As you approach each of these experiments, keep in mind and discuss the Scientific Method—that baseline for scientific inquiries used for hundreds of years to explore and explain universal phenomena. The Scientific Method tells you and your lab partner to:

1. Observe an activity or phenomenon.

2. Come up with a possible explanation—your *hypothesis.*

3. Use your hypothesis to make predictions about the activity or phenomenon.

4. Experiment! Test your predictions. Get your hands dirty.

5. Come to a conclusion about your hypothesis and its ability to predict the activity.

You probably use the Scientific Method in everyday life more than you realize. You make breakfast and realize that the toast you put in the toaster hasn't popped up yet and it's been a while—too long, in fact. You slap your head and think, "I bet I didn't plug in the toaster!" This is your hypothesis. You go check it out and see that, sure enough, you didn't plug in the toaster. So you do. This is your experiment. And sure enough, the toast pops up in a matter of minutes. (Then you slap your head again and come to your conclusion: You need to drink more coffee.)

Aside from helping us out of breakfast jams, so to speak, the Scientific Method helps us explore the universe around us and find explanations for the things we see. It's a critical component for testing our guesses and finding answers to them. At a fundamental level, it also helps build critical-thinking skills for young students, so that when they're faced with those who would laugh in the face of science, they can look back with a steady eye and say, "Prove it. Show me your experiments and conclusions."

The Scientific Method gives us a common way to make guesses about our observations, test them out, and see if we're correct. Others should be able to replicate your experiments and come to the same conclusions. At its heart, the Scientific Method is a system of critical thinking and analysis that will allow your lab partners to think for themselves and not be dissuaded from sound science in the face of emotional arguments or ancient fictions.

Make Way for Science!

The bottom line is that all the experiments you'll find throughout this book will give you and your kids a way to have a great time together—whether you're exploring how balance works or launching a soda rocket in your backyard. The experiments and the time you spend together doing them can also go a long way toward building a foundation of knowledge and an understanding of the world, how it works, and how your kids can begin to answer their own everyday questions.

So break out the kitchen supplies, roll up your sleeves, and make way for science. Madly, impossibly awesome science.

Exploding pockets of steam. Water that is heavier than *other* water. Metal that *falls apart*. Chemistry, or the exploration of matter, always makes me feel like a mad scientist.

In a good way.

In the experiments in this chapter, you'll learn how crystals are formed, how metal can be bent with simple household chemicals, and so much more.

Maybe you'll even learn a few things they didn't teach you in school. For instance, you may remember your schoolhouse teachings about the three states of matter: solids, liquids, gases. But for some reason, our culture takes a long time to catch up to science, which boasts *four* states of matter—plasmas being added to the mix. Considering how abundant plasma is—it's in everything from fluorescent light bulbs to the stars you see at night—it's a little odd that plasma is frequently left out of the matter list, but if you didn't learn about it when you were in school, you'll learn about it soon enough as we explore the amazing, mad scientist powers of chemistry.

SOAP CLOUDS

A childhood memory came back to me while doing this experiment. Like any kid, I loved bath time—the bubbles, the toys, the warm water. It was like a playground at home, in the bathroom. Eventually, however, every playtime is interrupted by a parent who comes in and demands that the bath be used for its intended purpose: cleaning.

That's right. Parents. We're killjoys now . . .

But back to the memory. Whenever my mom came into the bathroom and tried to hurry along the process, I would have to try to find the soap—that slippery little worm that somehow vanished and squirted about this way and that. It was impossible to find the bar of soap in the sudsy water.

Until, that is, we discovered Ivory Soap, the one white bar that floated on the surface. Turns out the company pumps air into the ingredients during the soap-making process—enough air to make the bar float when it lands in your bathtub. This is key, because in the upcoming experiment, you'll learn how to use your microwave to make all that air explode and expand creating great, puffy clouds of soap that help you learn about steam power and also help make bath time even cooler.

So dig around your medicine cabinet or pick up a bar of Ivory Soap the next time you're at the store and try this little experiment. It's really quite simple and fun. All you do is throw it in the microwave for a few minutes and watch as the air and water inside the bar expand the whole thing into soap clouds. The cool thing is that afterward you can just toss it into the bath and let your soapy little lab partner go bonkers.

Fun time in the bath, guaranteed.

WHY IT WORKS

A bar of Ivory Soap has tiny pockets of air that allow it to float on the surface of the water. Well, those air bubbles also contain trace bits of water—just enough to vaporize (or turn to steam) and expand when heated in the microwave. Combine that with the softening of the soap, and all that heated water keeps searching for places to expand, making the soap grow and grow and grow, until you have a great big fluffy cloud of happy happy fun time.

Seriously. This one is a quick and easy winner in terms of fun, but it also helps explain what happens to gases when heated. They expand! Think about the popcorn you cook in the microwave. The same principles are at work. When you make popcorn, water in the unpopped kernel heats up, is vaporized into steam, and is soon on the lookout for a place to keep expanding. Eventually, the outside kernel can't hold in the heat and voilà—snack time.

The expansion of the popcorn kernel and the soap shows Charles's Law in action, or the law stating that as the temperature of a gas increases, so does its volume. Jacques Charles was a scientist, mathematician, and . . . wait for it . . . hot-air balloonist! No wonder he was interested in how heated gases expand.

 HERE'S WHAT YOU NEED

- ❑ Bar of Ivory Soap
- ❑ Plate and paper towels
- ❑ Microwave

HERE'S WHAT YOU DO

1 All you really have to do here is have your lab partner unwrap the bar of soap, place it on the plate, and then microwave it for 2 minutes. I wouldn't go more than 2 minutes, but microwaves vary, so keep an eye on what happens. Besides, it's fun to watch the cloud expand! It sort of starts out slow and then begins to build into a puffy, meringue-ish ball about triple the size of the bar.

2 The Ivory Soap is still soap. Nothing has happened to it except expansion. So, at bath time, just chuck your new soap cloud in the water and let your mad scientist cackle with glee.

W ANT MORE?

If, by chance, you don't have any soap, the same principles can be seen with marshmallows. Put a few on a plate and watch them expand in the microwave, but keep an eye on them because they expand really quickly. You may want to prepare a cup of hot cocoa and grab a book before you do this one. If you overheat the marshmallow a tad, you'll end up with the most delicious caramelized science treat you can ask for.

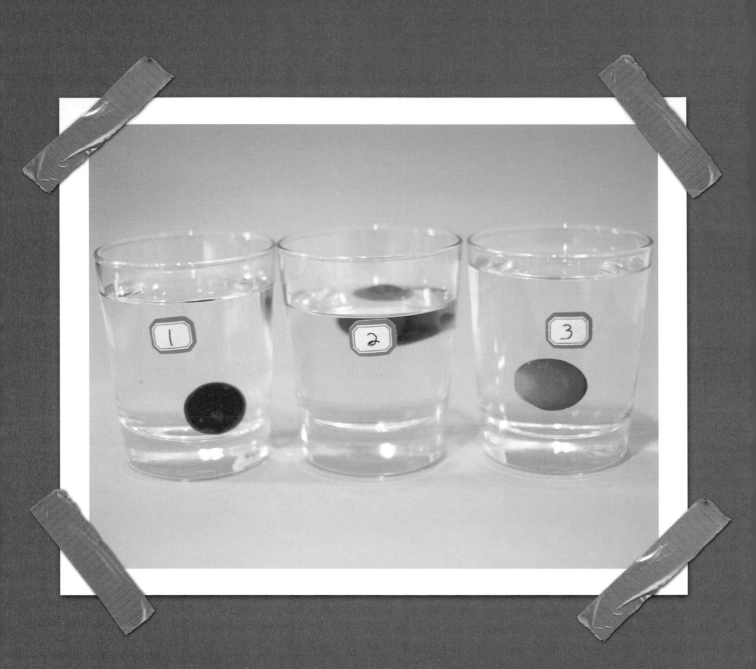

FLOATING GRAPE

This is a fun experiment that delves into density, or how solid something is.

Have you ever noticed that ice floats in a glass?

Well, ice is less dense than water. It is lighter. So it floats.

But how about different objects in the same glass of water? How about, say, a grape? What do you think will happen to it? Will it sink? Float? Sprout wings and fly away?

In this experiment, you and your lab partner will be tinkering with liquids to make them more dense than a grape, so that the grape can float instead of sink. Then you'll do something extra nifty with the grape and put density experiments to the test. Oh, fine. I get it. The suspense is killing you. So here's the deal: You're going to float one grape to the top of a glass, sink another, and then make one appear to magically float in the middle of a glass. Nifty, eh?

WHY IT WORKS

Density is the measurement of how solid something is. In the case of glass Number 1, which is filled with plain water, the grape is more dense or solid than the water, so it sinks. But with glass Number 2, which is filled with a homemade sugar solution that is denser than the grape, so the liquid actually supports the grape and allows it to float.

Ahh, but glass Number 3. There's the catch. The bottom will be filled with your sugar solution but the rest of the glass will be topped off with plain, sugar-free water. The grape sinks through the fresh water because it is *more* dense than the water. But it is *less* dense than the layer of sugar water on the bottom, so it appears to float in the middle of your glass—caught between liquids of varying densities.

W ANT MORE?

Try floating grapes in different liquids you might already have in the house, like fruit juice, sparkling water, cocoa, or coffee. Or try to come up with a solid sugar-water solution that will allow the grape to float in the middle just like it does in glass Number 3.

HERE'S WHAT YOU NEED

- ❏ Three drinking glasses
- ❏ Measuring cup
- ❏ Water
- ❏ Sugar
- ❏ Tape and pencil to mark glasses
- ❏ Grapes
- ❏ A spoon

HERE'S WHAT YOU DO

1 Have your lab partner mark all your drinking glasses, 1 through 3.

2 Let your lab partner fill your measuring cup with water and stir in sugar until you can float a grape easily in the liquid. Take your time and find out just how much sugar this requires in, say, 2 cups of water. Or you can just stir in as much sugar as humanly possible. Just make sure your mix is dense enough to float a grape. Again, this is super easy and messy and finger-licking fun, so let your lab partner take the lead while you're at the ready to explain density.

3 Now that you have your mix, turn your attention to glass Number 1. Fill it with fresh water, not the sugar solution, and then add a grape. What happens? It sinks, right?

4 Now it's on to glass Number 2. Fill it with half or so of your sugar solution and add a grape. The grape floats, right? *(See Fig. 1.)* You should know this already from making the solution . . . unless you're, wait for it, dense.

5 Now here's the cool/tricky part. Take glass Number 3 and fill it *half full* with your sugar solution. Then, take your spoon and place it just above the top of the liquid and then fill the rest of the glass very, very slowly with water, letting the water dribble off the back of the spoon into the mix, being sure not to mix the two different liquids. In other words, the water should float atop your sugar mix. Don't stress out if it takes a few tries to get it right. Now take a grape and gently drop it into glass Number 3. If all goes well, the grape should float in the middle of the glass, between your liquids of varying densities. If it doesn't, just lick the sugar off your fingers and try again.

Fig. 1

PENNY SHINERS

Did you know that when the Statue of Liberty was assembled in New York in 1886, it looked pretty much like one of the dull brown pennies in your pocket or piggy bank? What we now know as a beautiful green statue was originally the same color as coins.

What happened? Here's the story.

Every metal undergoes a change when it encounters oxygen.

In the case of the Statue of Liberty, there's a cool exchange of chemicals going on. The brown copper mixes with oxygen and suddenly you have copper oxide. Mix that with the chlorine from salt from the surrounding sea and boom, you have a patina, or covering, of green. It took years and years—decades in fact—for the entire statue to turn the shade of green it remains today.

You can see the process in action using just a few pennies and some chemicals found in your kitchen. A simple soak in regular vinegar and salt is enough to clean off dirty pennies, but then watch what happens as you remove the pennies and let them air dry. The salt and air will turn them splotchy green if they're not rinsed off after the bath.

HERE'S WHY IT WORKS

Over time, pennies, which are made with copper, become coated in copper oxide—which is formed when copper and oxygen mix together. It's the copper oxide that dulls the shine of the copper and makes them look so dirty.

The good news is you can remove the copper oxide with acid and vinegar and salt (which you'll use in the experiment), combined to make the perfect weak acid for the job. How very convenient. You'll notice the acid stripping away the copper oxide and making the penny shiny again almost right away when you dip the pennies halfway into the liquid. But you can also keep going with this experiment after a good penny bath.

After 5 minutes of soaking, you'll take the pennies out of the acid. You'll have one set that you'll rinse off and one that you won't rinse. You'll place both on a paper towel. What happens is pretty cool but it takes a little more time to see. The pennies you rinsed off? They're basically done. They should remain shiny for a while. The other set of pennies, however, will undergo another chemical change. Now that you've removed the copper oxide, it should be easier for the copper on the penny to combine with oxygen *and* the chlorine left over from the salt bath—creating a new, bluish-green compound known as malachite. Your pennies should have little speckles of blue-green malachite after a few days.

 **W** ANT MORE?

Try leaving a few pennies in the vinegar-salt acid solution for much longer, just to see what would happen. You'll be surprised at the results. One penny might get noticeably shinier, while the other might break apart. For us, it got to the point, about a month into the process, that we could flake off the edges and even *bend* the penny, because the acid eroded, or ate away, the metal.

HERE'S WHAT YOU NEED

- ❏ ¼ cup vinegar
- ❏ 1 teaspoon salt
- ❏ A glass
- ❏ Pennies! The dirtier the better
- ❏ Paper towels

HERE'S WHAT YOU DO

1 Mix the salt into the vinegar until it is dissolved. That's the hard part.

2 Have your lab partner dip half of a penny into your solution. What happens? Is it suddenly shinier? Keep a very close eye on the submerged half. Often you can see the changes right away. Take it out after 10 seconds or longer to note what happened.

3 Now drop all your pennies into the mix and take note of what happens. Do you see anything? Keep a close eye out for quick, chemical cleanliness. *(See Fig. 1.)*

4 After 5 minutes, take out *half* of your pennies and rinse them off. Place them on one paper towel. Now, take out the rest of your pennies and *don't* rinse them. Place them on a different paper towel.

5 Take note of the two different groups. Let them vegetate for a few days. One set should remain shinier than they were when you began, while the other set should start to take on splotchy green areas.

Fig. 1

BOILING ICE

Have you ever boiled some water for a quick, midweek pasta dinner only to toss in the pasta and watch the water stop boiling immediately? This always frustrates me, because I keep thinking, "But the package says *rolling* boil! And this isn't even a boil at all!" Ahh, but this is physics at work. In this experiment, you will see not only how water can turn to gas, or steam, but also how heat and energy interact with the different states of matter. You'll basically be boiling a big pot of water and then throwing some ice into the mix. Watch what happens. The water should stop boiling for a bit and then start up again. It's an amazingly simple experiment using just a few kitchen ingredients, but there's a lot to talk about. Who says you need that fancy science kit for cool experiments? You don't. You just need a little time before dinner . . .

HERE'S WHY IT WORKS

Thermodynamics, yo. That's why. Boom. Lesson learned.

Okay, okay. There's more. And believe me, *thermodynamics* is a fancy word your little scientist is going to come across again and again in later physics and natural science classes when she's older. So this is a great starting-off point to discuss a key part of how our universe works.

In this experiment, you'll boil water and drop in an ice cube. When this is done, suddenly there is a cool spot in the pot. The heat stops boiling the water and instead focuses its energy on the cool spot, flowing toward the ice. So now you've got a pot of hot water filled with ice. What happens next? The ice melts and then? That's right. There aren't more cold spots so the water returns to a boil.

In short, the ice stops the boiling because the Second Law of Thermodynamics—the idea that heat will always flow to the coolest point—is at play. Think about what happens when you add a spoon to, say, a cup of coffee or hot chocolate. Leave it in for a while and the heated liquid will seek out the cool spot, the spoon, and warm it up. Cold pasta. Boiling water. Same deal.

This is a super-easy experiment to set up a discussion about how heat and energy work on a fundamental level. But there's more at work here as well. You started off with a liquid in the water. But then, after the boiling began, you probably noticed steam. Liquid and gas are two of the four fundamental states of matter—with solids and plasmas rounding out the group.

⇨ HERE'S WHAT YOU NEED

❏ Pot of water

❏ Stove

❏ Ice—not much, a few chunks. Cubes. Whatever you want to call them. (Note: To better see what's going on in the pot, use water tinted with food coloring to make the ice.)

HERE'S WHAT YOU DO

1 Boil water. You can do it!

2 When it's really roiling, have your lab assistant drop in a few ice cubes and watch the water stop boiling and then start up again. You'll be able to see the dye from your ice cubes spread out into the water as the ice melts. Seriously, that's it. Yes, it's simple and easy—too easy, really. But what's at play here are the building blocks of explaining thermodynamics and pre-engineering life lessons.

DID YOU KNOW?

Speaking of matter. There are four fundamental states of it:

- **Solids** are hard and usually dense and will hold their shape outside of a container.

- **Liquids** are materials that are fluid—less dense than solids but more dense than gases—and will form to the shape of a container but won't expand to do so.

- **Gases** are high-energy states made up of randomly moving, high-speed molecules that will conform to the shape of a container and will expand or compress to do so.

- **Plasmas** are superheated gases that often shine very brightly. Stars, lightning balls, the Northern Lights, neon lights: These are examples of gases that have been heated so much that their internal structure changes and they become plasmas.

ROCK CANDY CRYSTALS

Full disclosure: We messed up with this one. We messed up with this one *gooooood*.

Not only did we—*I*—mess it up, but we invited over a friend, got her superduper excited about making rock candy, and even gave her extra rock candy sticks and jars of goo mix to share with her brother and sister.

"It should be ready in a week!" we told her, "Have fun!"

Then we sent her merrily on her way with a rock candy kit that was just absolutely, completely messed up.

I imagined her checking her rock candy stick every day for a week, just waiting for it to work—maybe, perhaps, telling her siblings how amazingly awesome it was going to be. Poor kid. She had no idea I had messed up this experiment.

But you know. Sometimes science is like that. So what do you do? You roll up your sleeves and experiment some *more*.

Rock candy making is usually one of the coolest ways to see the formation of crystals in action. It's really quite simple: You boil some sugar and water, add some flavoring or some food dye—or heck, *both*. And then you dump it in a jar, add a sugar-coated stick for the rock candy to form on, and then a week or so later there it is: pure sugar joy. It's the process of molecules clinging together in action. It's also quite delicious.

Just make sure not to accidentally scrape the sugar off the stick when inserting it into your jar, because then you'll have nothing for the sugar crystals to form on—and you'll also have a pack of upset seven-year-olds who think you enjoy teasing children. Not. Fun.

So here's what will hopefully be a surefire way to make rock candy, see crystals in action, and not upset all the neighborhood kids.

HERE'S WHY IT WORKS

In short, the sugar molecules stick to each other, growing and growing as more and more molecules stick together.

Now for the long explanation: With the sugar water, you have created a saturated solution, which basically means the liquid (water) can't hold any more compound (sugar). The molecules, a mashup of atoms that make sugar sugar, are crowded around in your solution and start to bump into each other, sometimes sticking together in a process called *nucleation*. During this process, the crystals form and grow bigger when more and more sugar molecules stick to each other. After just a day or so, the growth is visible. After a week, it should be big enough to look mighty tempting. I don't know what it looks like after two weeks, because we can never resist eating it for that long.

By adding a sugarcoated string or stick to your solution, you kick-start the nucleation process by giving the molecules a great place to latch onto. (Remember how we messed this one up? By scraping the sugar off the stick, we didn't give the sugar molecules a place to latch onto in order to kick-start crystallization. Instead, all the crystals started forming on the bottom of our jar, where all the starter crystals had fallen. Hopefully you will learn from our mistakes. You're welcome.)

If you'd like, set up two experiments. In one, sugarcoat a stick before adding it into the solution, and in the other try *not* sugarcoating a stick. See what happens. Just be prepared to be the one scraping sugar crystals off the bottom of the jar while your kid's snacking on some delicious rock candy.

HERE'S WHAT YOU NEED

- ❏ 4 cups water
- ❏ 2 cups sugar, plus a little extra to coat sticks or pipe cleaners
- ❏ Saucepan
- ❏ Mason jar—small, big, whatever you have
- ❏ Food dye and/or flavor oils (optional)
- ❏ Long wooden toothpick and/or pipe cleaners or butcher's twine
- ❏ Pencil or more wooden toothpicks
- ❏ String for tying, if needed

HERE'S WHAT YOU DO

1 Heat up your water to a light boil and then add sugar a bit at a time until it is dissolved in the mix before adding more. Keep adding more until all your sugar dissolves in the boiling syrup and you have a nice goopy soup of sugar. Then let the mixture come to a rolling boil for a little bit—maybe 30 seconds.

Note: I usually don't preach adult supervision because kids can, more often than not, do stuff a lot better than adults anyway and how are they going to learn if they're not allowed to make mistakes? But boiling sugar absolutely scares the bejesus out of me. If that stuff gets on anyone, it's not only going to burn, it's going to stick and burn. And that just sucks beyond words. So there you go. A warning. Be careful.

2 Once the mixture has boiled for 30 seconds, remove from heat and add food dye or flavor oil if you'd like. A few drops of dye or a capful of vanilla or spearmint should do the trick quite nicely.

3 Let the mixture cool, then have your lab partner add it to the Mason jar, and set it aside to cool some more.

4 While the mixture is cooling, wet your stick or pipe cleaners and then sprinkle with sugar. If you use butcher's twine, dip it into your newly created sugar mix and then remove to dry for a day. This gives a rough surface for the sugar crystals to latch into and begin formation. It's like seeding them for crystal growth.

5 Once everything is relatively cool—it doesn't have to be room temperature; warm will do—add your stick or cleaner to the mix. You can tie the pipe cleaner to a pencil and then lay the pencil over the mouth of the jar, so that the pipe cleaner doesn't touch the bottom of your jar. (This is where we messed up, because we put the Mason jar tin lid back on, poked some holes in it, and then tried to force our sugar sticks through the holes. All the sugar scraped off and there was nothing for crystals to form onto. Oh well. Keep Calm and Science On.)

6 Now the waiting game begins. You should start to see some good crystallization in a day or so. Store in a dark space for a week and keep track of the crystal growth. Feel free to let the crystals grow as big as you'd like. But a solid week usually does the trick. When you're ready, remove the cleaner or stick from the mix and let dry on a plate or in another jar for a few hours to dry. Then, inspect and enjoy. *Crunch, crunch, crunch.*

 DID YOU KNOW?

People all over the world and throughout history have used rock candy for medicinal purposes, mixing herbs or alcohol in the solution to create throat lozenges, for instance.

Fig. 1

RAINBOW WATER STACKS

Look, I'll level with you: Do this experiment one way and you might just have as difficult a time as we always do. But do it a different way, and you have a delicious cold treat! So I'm including a few different options to experiment with density, or the measure of how solid something is.

With the first, most challenging experiment, the idea is to "stack" a rainbow of water, using just sugar to alter the density of water and make different batches of colored water float on top of each other. No matter how much we tinker and experiment with our rainbow, we just can't seem to pull it off in a way that blows the mind. Some colors stack and some colors blend. As it turns out, this is a good thing—because while I always get a little annoyed (I want satisfaction *now*!), my little scientist absolutely *loves* the challenge. "That's just what science is!" she says, "You have to fail and try again!" It warms my cold, scientific heart . . . even if it also wonders, "Why can't I make this work! How am I not smarter than water?!"

I mention all this because experiments don't always work out as you planned, so don't worry about it. Have fun. Take a lesson from your little scientist and keep working!

All told, it's a fun, messy, hands-on experiment that helps explain density. But if the rainbow water experiment doesn't float your boat, there are many, *many* different ways to discuss density.

Here, you'll find three experiment options: two that are either for looks or drinking, and one that is just plain poison, so please, don't drink the one with bleach. But please note, the one with bleach is downright flippin' sweet! It's *sooo* much more fun than any of the others, because the bleach's chemical properties will create a pretty cool reaction that is instantly observable.

So there you go. Three experiments in one. Try them all!

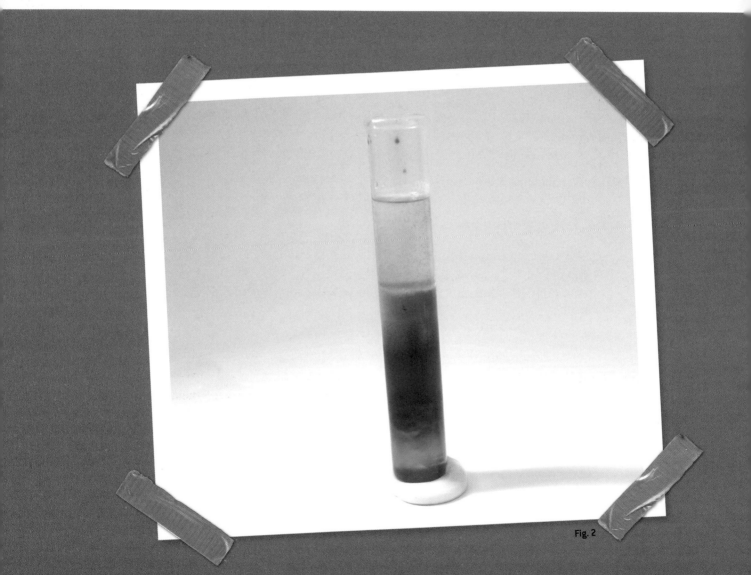

Fig. 2

WHY IT WORKS

Have you ever noticed that ice floats in a glass of water? Or that when you throw a rock into a lake, the rock sinks to the bottom? Remember the floating grape experiment? It's all about density. Ice is *less* dense than water, so it floats. A rock is *more* dense than water, so it sinks. In all these experiment options, the liquid that is most dense will sink to the bottom of your glass, while the liquid that is less dense will float on top.

In a dense liquid, all the particles that make up the liquid are packed together tightly, forming a liquid that is more solid and capable of supporting liquids that are less dense, or that have particles that are more loosely packed together. Think about it this way: If you put a Ping-Pong ball on the ground and then put a book on top of it, the book is just going to slip off or tilt, right? Now put a dozen Ping-Pong balls close together on the ground and try resting the book on top again. The book is supported, right?

In Option #1, the sugar water stacks should float on top of each other, because the densest water, the one with the most sugar, sinks to the bottom while the least dense water, the one without any sugar at all, floats on top of all the other layers. If your rainbow stacks don't look quite like a rainbow but rather resemble an oozing brown sludge (don't worry; it happens), try tinkering with how much sugar you add to each glass. And also make sure you pour the liquids in super, super slowly. Give each layer a few minutes to settle before adding the next color.

In Option #2, when you add bleach, it will sink below one layer and then float atop another. Take special note of what happens when you add the bleach to the mix. The bleach should pass through the oil layer, because it is more dense than the oil. But because it is less dense than the corn syrup, the bleach should remain with the water in the middle, where it will mix with the blue water and turn it clear again. Cool, right? Again, just to be clear, *don't drink this one*!

In Option #3, the placement of the floating liquids all has to do with how much sugar is in each drink. The red layer, on the bottom, should have the most sugar and be the liquid with the highest density. The blue Gatorade should be the second densest liquid, while the clear diet soda has the least amount of sugar and should be the least dense of all. Pouring over ice helps keep the liquids spread out when pouring.

 HERE'S WHAT YOU NEED

OPTION #1: RAINBOW WATER STACKS

- ❏ A clear bottle. Or vase. Or clear tube of some sort. You get the idea: something big enough to hold a lot of liquid.
- ❏ Four glasses
- ❏ Sugar (Lots!)
- ❏ Food dye
- ❏ Water
- ❏ Measuring spoons and glass

OPTION #2: COOL BLEACH LAVA LAMP *(See Fig. 3)*

- ❏ Red and blue food dye
- ❏ Clear bottle
- ❏ 1 cup corn syrup
- ❏ 1 cup vegetable oil
- ❏ ½ cup water
- ❏ ½ cup bleach (This is why we don't drink this one!)
- ❏ Measuring glass

OPTION #3: SWEET, DELICIOUS DENSITY TREAT *(See Fig. 4)*

❑ Ice! Ice is key. Use so much ice it hurts your brain.

❑ Clear tall drinking glass

❑ Cranberry juice cocktail (or super-sugary red fruit punch)

❑ Blue Gatorade

❑ Diet 7Up soda or Sprite (Make sure it's diet!) Or you can just use sugar-free sparkling water.

Fig. 3

HERE'S WHAT YOU DO

1 **Option #1:** Rainbow Water Stacks: Your lab partner can pretty much do all this. It's just a lot of measuring and pouring. Take your four glasses and have your lab partner fill each of them with ½ cup water. Now, dye one glass of water green and add 6 tablespoons of sugar, stirring until it is dissolved and then set aside. Take your second glass and dye it blue, adding 4 tablespoons of sugar and stirring. Take your third glass and dye it red, adding two tablespoons of sugar. Take your last glass, dye it yellow, and don't add any sugar. Now pour the green water (the water with the most amount of sugar) into your bottle first. Then, take your blue glass of water—the one with 4 tablespoons of sugar—and very, very slowly pour the water into your bottle, being sure the blue water touches the side of your bottle. If you'd like, you can pour over a spoon as well to help spread out the liquid. Repeat the same process, first with your red water and then with the yellow water. At this point, you should have a miniature, muddled-looking rainbow of color in your bottle, as the water floats on top of the color below it. *(See Fig. 2.)* If, however, you're having trouble with

this, try tinkering with the amount of sugar you use. More sugar will create denser liquid, so try messing with the amounts to make a rainbow stack to your liking.

2 **Option #2:** Bleach Lava Lamp: Mix red food dye with the corn syrup and pour into the bottle. Then pour the oil into the bottle on top of your syrup. The oil should float on top of the corn syrup, because it's less dense. Now add the blue food dye to the water in your measuring cup. Pour the water on top of the oil in your bottle and give it a few minutes to settle. It will actually sink below the oil and rest atop the corn syrup, creating a middle blue level. Once the water has settled, you should have three distinct layers of liquid at this point. The bottom should be red, the middle should be blue, and the top layer, the oil, should be clearish. Now pour in the bleach and watch what happens as the bleach eats away at the coloring—especially to the blue portion of your bottle. *(See Fig. 1.)* It takes only a few seconds for chemical reactions to occur, so keep observing as the blue layer completely disappears. I always think this experiment resembles a nifty Lava Lamp.

3 **Option #3:** Sweet, Delicious Density Treat: First things first. Fill your glass with ice. And when I say fill, I mean *fill*! Pour just enough red fruit punch or cranberry juice to create a bottom layer in your glass. Don't fill up to halfway. Now, very, very slowly, pour in your blue Gatorade over the ice. You should notice the blue layer appears to float atop the red. Wait a tiny bit for things to settle. Now take your clear diet soda or sparkling water and pour *super* slowly over the ice. It should float at the top. *(See Fig. 5.)* Enjoy your red, white, and blue sugar buzz!

Fig. 4

 **W ANT MORE?**

While the drink created in Option #3 makes for the perfect Fourth of July picnic punch, try other combinations of beverages and colors for different events.

Fig. 5

Biology, or the study of life, is a wide and varied field of science that provides for endless, fascinating explorations into what makes the world tick, into what makes *us* tick.

Take the time to look around for a second. What do you see, hear? You've probably seen your little scientist absorbed in this very act a million times over. It can be frustrating, believe me, when you're trying to keep a schedule and there's your little lab partner, bent over a flower and staring in awe. But on a small scale, this is biology in action. You get to get your hands dirty.

In the experiments in this chapter, you and your little lab partner will take a deeper journey into biology and find answers like what makes plants thrive (Colored Leaves), how do animals adapt to their environment (Animal Camouflage), and why worms come out at dusk (Light Fright). Biology seeks to answer these and millions of questions just like them through the study of the world around us. So get ready to roll up your sleeves and get dirty.

COLORED LEAVES

You've watered plants before, right? You know how it works: You plant something, give it some water, give it some light, give it some time and voilà! Beans! Or broccoli! Or flowers!

But . . . *how* does it work?

How does the water go from the dirt into the plant and on up into the flower?

Next time you have some beautiful flowers in a vase at home, observe what happens to the water in the vase over a few days. It disappears, right? Well, where is all that water *going*?

This is a fun experiment that will detail in full, beautiful color how everyday plants receive and process the life-giving water they need to thrive. With a few glasses of colored water, some flowers, and some cutting, you'll watch as flowers drink up their nourishing water and display new and brilliant colors.

HERE'S WHY IT WORKS

When flowers "drink" water, what you're seeing is called *capillary action*, or *capillarity*. That's the ability of a liquid to rise through narrow spaces—in this case, a stem. Although gravity should force the water down, capillary action allows the water to rise, first through the stem and then up, up, up into the flower petals. In this experiment, notice how the petals begin to take on the same color as the dye you've added to the water? That's because the colored water rises through the flower stems. You can literally watch capillarity in action—although it is, indeed, a slow process.

 ## HERE'S WHAT YOU NEED

❏ Three white flowers. We used Gerbera daisies, but you can use anything you have on hand or that is easily available at the florist or store: roses, carnations. Go pick some lawn daisies if you have them. If you don't have flowers, celery works great as well.

❏ Four glasses of water

❏ Food dye: blue, yellow, green, red

❏ Sharp knife

HERE'S WHAT YOU DO

1 Let your lab partner mix each color into each of your glasses, until you have four glasses of different colored water. The stronger you make the colors, the better. *(See Fig. 1.)*

Fig. 1

2 Add two flowers to two colors of your partner's choice and set aside. You should have two colors and one flower remaining at this point. *(See Fig. 2.)*

Fig. 2

3 Take your last flower and have your lab partner slit the stem *(see Fig. 3)* so that you'll be able to put one part of the stem in one remaining colored glass and the other part of the stem in the other remaining colored glass. *(See Fig. 4.)* In other words, your flower should now be drinking from two glasses at the same time. Got it? Good. Note: Let your lab partners do this if you feel they're up to it. Even the cutting. Teach tool safety and step aside if they're comfortable with it.

Fig. 3

4 Place the water glasses away from the sunlight and watch what happens. You'll start to notice some fascinating results within an hour or so, as the colored water creeps up the stem and begins to color the leaves. But wait a few days and keep observing the petals for fuller color displays.

Fig. 4

 DID YOU KNOW?

Xylem is the name for one type of plant tissue that carries water. It comes from the Greek word *xylon*, meaning "wood." The other tissue is called phloem. It comes from the Greek word *phloios* meaning "bark." Together, these tissues form the main capillary action that helps plants drink water and nutrients and survive.

LIGHT FRIGHT

Whether you realize it or not, and whether you have a family pet or not, your yard is full of poo.

Worm poo.

And the nutrients in the poo are part of what keeps your grass and plants thriving and looking so wonderful. (And also what keeps us all breathing clean air, as those thriving green plants convert waste gases to oxygen!)

Worms gobble up soil filled with decaying plant matter and then, when they're done digesting it, excrete what worm enthusiasts call *castings*. The castings are nutrient rich and help make it easier for other plants to grow. Much the way plants take carbon dioxide waste from the air and turn it into breathable oxygen, worms take wastes in the soil and help make the dirt fertile again. Thanks for the help, worms!

Plus, here's a cool thing: Worms have five *aortic arches*, or blood-pumping vessels we know as worm hearts, and they breathe through their *skin*.

Amazing, amazing creatures. But how do they function? How do they . . . *live*? And if they're all over the yard, helping us out so much, why can't we *see* them? In the following experiment, we'll explore the question of why worms only seem to come out after heavy rains. In short, that's because they hate the light and can only breathe in wet environments. But let's explore some more.

WHY IT WORKS

In this experiment, you and your lab partner are going to see whether earthworms prefer light or darkness. Ask your little scientist and make a hypothesis about worm preference. (Hint: It's the dark.)

But before we go any further, let me tell you this: Earthworms don't have eyes. But they *do* have light-sensing spots near the top of their heads. They use these to sense whether it's dark enough outside to emerge from the soil into the nighttime air.

This is important because worms—which don't have lungs—breathe through their skin and they need moisture to help with the process. If it's light out, it's probably daytime and therefore possibly hot. So they're going to stay underground where it's cool and moist. But at night, their sensors will tell them it's probably okay to come out because the moisture in the air won't evaporate so quickly. In other words, they'll come out at night when they know they can breathe better.

Isn't that funny? If humans dove underground, we'd have to hold our breath. Worms, on the other hand, wouldn't be able to breathe *aboveground*—at least in hot, dry environments. You've probably seen worms scurrying about on the sidewalk on rainy days, right? That's because darker, cooler air + moisture = easy breathing!

 DID YOU KNOW?

The earthworms you found can grow from 6 to 11 inches long. Giant worms in Australia and South America can reach 11 feet in length!

HERE'S WHAT YOU NEED

- ❏ A shoebox

- ❏ Scissors

- ❏ Paper towels

- ❏ Earthworms (Your lab partner can probably find a few with a little digging—just be careful not to hurt them! The worms, that is. Your lab partner is tough.)

- ❏ Desk lamp (or flashlight and tape!)

HERE'S WHAT YOU DO

1 Take your shoebox lid off and cut off about one-third of it. Go on. Just cut it off. This will be the perfect gig for your little lab partner.

2 Now wet several pieces of paper towels and put them in the shoebox, creating a nice ground cover. Now gently place your worms—let's shoot for at least three or four—on one side of your shoebox. *(See Fig. 1.)* Be gentle with these guys—and girls. Earthworms are *hermaphrodites*, which means they have the reproductive organs of both genders.

3 Now put the lid back on the shoebox, with the opening on the same side as your little worm collection.

4 Put the open end of the shoebox under your desk lamp and turn on the light. Don't put the light super close to the shoebox. You don't want to burn the worms. Just make sure the light is on them.

5 Now leave the box for 20 to 30 minutes and see what happens. Where did the worms end up? Did they stay in the light? Move to the other side? Burrow under the paper towels? They should have moved away from the light and even have dived for cover under the towels.

Fig. 1

FALLING LEAVES

You've probably taken a hike and fielded these questions: Why are the leaves green? Why are they yellow? Wait a minute, now they're brown? How does *that* work?

Outdoors + children = some of the best questions and times that can be had.

And of course you respond that in the summer, leaves are green and lush. But every fall, like clockwork, the leaves on some trees turn from green to beautiful shades of yellow, brown, and red. Eventually, all those leaves fall to the ground, creating hours of raking chores and, if you're lucky, hours of leaf-jumping fun (or not lucky, depending on who is handling the rake).

But what's at work here? What makes the trees around you turn color?

If you have five minutes to set this up—and then several hours of waiting time to spare—you can create your own leaf-peeping tour right in your own kitchen, using a few common household supplies. This is the perfect experiment to welcome the fall season.

WHY IT WORKS

In this experiment, your lab partner will mash up some spinach leaves, add some nail polish remover, and then sit back and watch as the chemical soup turns a bit of coffee filter green, yellow, red, brown—all the colors of the fall season over time.

But . . . how?

At work here is the chemical reaction that keeps everyone on earth alive.

You've heard of *photosynthesis* before, I'm sure. It's the process of plants turning sunlight, water, and carbon dioxide into their food, a process that also releases oxygen into the air for the rest us to breathe. Yay!

The chemical in the leaves at the heart of this process is called *chlorophyll*, a compound that helps convert light into energy, or plant food. In the sunny parts of the year, there's enough light to keep that process moving along, and leaves are green and lush.

But what happens in the fall? The hours of sunlight grow shorter and shorter, and so trees produce less and less food and the chlorophyll starts to disappear. Without chlorophyll, the green tint of the leaves fades and you're left with an earth-toned rainbow for a few weeks until the leaves begin to fall.

In this experiment, you'll notice different colors on your coffee filter as the chemicals in your spinach mash are released and the chlorophyll decays.

 DID YOU KNOW?

Chlorophyll absorbs red and blue light waves, so that when you look at a tree or a leaf, all you see are the green leaves! Get rid of the chlorophyll and you see other light waves at work.

HERE'S WHAT YOU NEED

- [ ] A few spinach leaves
- [ ] A glass
- [ ] Spoon
- [ ] Nail polish remover
- [ ] White coffee filter
- [ ] Scissors
- [ ] Tape
- [ ] Pencil

HERE'S WHAT YOU DO

1 Have your lab partner break the spinach leaves into bits, put them in the bottom of your glass, and then mash them up with your spoon. *(See Fig. 1.)*

2 Let your partner add a few teaspoons of nail polish remover to the mix so that all the leaves are covered. Do not mix! Let the leaves settle while you move on to the next part.

3 Cut a long rectangular piece of coffee filter—narrow enough to fit into your glass—and tape the piece of filter to the middle of your pencil so that the filter dangles off of it.

4 Now, dip the filter into your glass so that the filter falls into the nail polish remover but is not long enough to reach the leaves at the bottom. (Don't worry if the paper touches leaves; not the end of the world.) The pencil will rest over the glass and keep everything from falling in.

Fig. 1

5 Let the filter sit for several hours and watch what happens. You'll notice changes relatively soon and then more and more as time goes on. Pretty soon, you'll see vibrant strips of yellow and green and, later, brown— a veritable leaf-peeping show in your own kitchen.

HOLE-Y WALLS

Plants and trees get the nutrients they need by sucking up water through their roots—a process known as capillary action (see Colored Leaves experiment in this chapter). But they also get nutrients through osmosis, a process in which liquid can pass through molecular walls. Imagine the leaves of a tree getting wet in a rainstorm. Through osmosis, the water and nutrients can pass through the leaf walls and keep the tree healthy.

But . . . how?

You may not realize it, but pretty much everything has holes in it—tiny, molecular gaps through which very small molecules can seep.

This experiment will show how leaves absorb water and will also show that supposedly water-tight plastic bags can actually absorb water through these tiny holes. This experiment is a fun one just for showing that sealed plastic bags are molecular Swiss cheese . . . if you work with small enough molecules. Learning about how plants remain alive is icing on the cake!

WHY IT WORKS

This is the perfect experiment to teach your little scientist all about osmosis.

Basically, you're going to watch as iodine—a mineral that helps regulate hormones and is stored mostly in your thyroid as well as in your stomach, blood, mammary glands, and even your spit, or saliva as it is known—sneaks *through* a plastic bag filled with a water-cornstarch mixture and colors the mixture inside. However, at the same time, the water-cornstarch mixture inside the bag isn't able to leak out and mix with the iodine. Why? Well, it all has to do with the size of the molecules of, well, everything. Iodine molecules are small enough to seep through the tiny, molecular-level holes in the plastic bag, but the molecules of the water-cornstarch mixture inside the bag are bigger than the tiny holes in the bag. This means that they can't get into the iodine solution. In other words, iodine can get *in*, but the cornstarch mixture can't get *out*. This is osmosis before your very eyes.

WANT MORE?

To experiment more with osmosis, try different materials for your iodine bag. Will it leak out of balloons? Latex gloves? Socks? Can the water-cornstarch mixture get into those new containers?

⇨ HERE'S WHAT YOU NEED

- ❑ Two big glasses
- ❑ Water
- ❑ Cornstarch
- ❑ A sealable sandwich bag
- ❑ Tincture of iodine (available at most drug stores for under $5)
- ❑ Tape

Fig. 1

HERE'S WHAT YOU DO

1 Have your lab partner fill both glasses with water. Three-quarters is perfect.

2 In one glass, she'll mix 2 teaspoons of iodine with the water. In the other glass, she'll mix 1 tablespoon of cornstarch with the water. *(See Fig. 1.)* Then pour about half of the solution in the plastic bag and seal it tight.

3 Now, rinse off the bag just to be sure there's no cornstarch on the outside of the bag. It will tinker with your results. (A clean science lab is an awesome, working science lab.) When you're good to go, place the sealed bag back in your iodine glass, being careful of overflow. Tape it to the side so the bag remains floating but not fully submerged. Keep the seal above the iodine, just so you can be sure nothing is leaking through the seal. *(See Fig. 2.)*

4 Allow the bag to stew for about an hour and keep track of your findings along the way. (You should start to see the first inkling of results after 5 minutes or so.) What's happening in the bag? In the glass? After a few minutes, you should start to see discoloration in the bag, a sign that iodine is getting inside. After more time, it should become inky gray. While you're waiting, add a few drops of iodine into your other glass, the one with the cornstarch solution. Take a look at how quickly the water changes color and keep that in mind when you then study the immersed bag. *(See Fig. 2.)*

Fig. 2

ANIMAL CAMOUFLAGE

My daughter and I were at the zoo when she was very young. We were watching all the animals you'd normally find on the African savanna: zebras, ostriches, gazelles.

"How?" my daughter asked, "Do zebras blend in with their surroundings when they stand out so much?"

Without giving it much thought, I started in on camouflage, or the ability of animals to disguise themselves. Emme cut me off.

"Dad, no, I understand camouflage. I just don't understand how *zebras* are camouflaged. I mean, *look at them*!"

And there they were—these black-and-white creatures against a green background. They were impossible to miss.

At home, we did some research on zebras and discovered that scientists believe zebras are meant to blend in with their herd, not necessarily the scenery. In other words, they blend in with *each other*. If a lioness launches an attack at a zebra herd, she might find it difficult to pick out an individual zebra among that zany mass of black and white.

We ended up talking for a long time about different animals and their camouflage, and how humans use camouflage as well. Soldiers who fight in the desert, for instance, have tan camouflage, whereas soldiers who fight in jungles use greens and browns, and soldiers in the snow use whites and grays.

Camouflage comes in all shapes, colors, and sizes, as your little lab partner will learn in this experiment.

WHY IT WORKS

In this experiment using simple arts-and-craft supplies, you can get a hands-on glimpse at how camouflage works by cutting up some squares of construction paper and seeing how different colors blend with different papers. A red square, for instance, will blend in on red construction paper. But a green square on the same red paper will stick out like a sore zebra. Or thumb. Whatever you want to call it. You get the idea. What's at play is a little bit of eye trickery and the science behind how some species survive and others . . . don't.

For instance, an animal that easily blends in with its environment can avoid detection by predators. Its species can survive, whereas an animal that sticks out might not last very long in the wild. (That said, there are also animals that *purposefully* stick out, such as some moths and butterflies whose wing patterns resemble predators. Some frogs make a show of themselves, alerting others that they're poisonous. "You can eat me if you want . . . but look at how colorful I am. I *must* be poisonous!")

There are several different forms of camouflage. *Crypsis* means blending in to surroundings. *Mimicry* means appearing to be something else, such as a bug that looks exactly like a stick. There's also something called *motion dazzle*, which is meant to confuse predators when the animal is in motion. Think about the zebra and its zany stripes.

With this experiment, crypsis is at play, as the squares of paper blend in with their surroundings. Take a look. It's a fun, simple game.

WANT MORE?

Try making a camouflage outfit by finding clothes that blend into the backgrounds found in your home. You can wear all white, for instance, against a white wall, or a floral shirt against some patterned wallpaper. You and your lab partner will have fun experimenting . . . and trying to sneak up on your family!

HERE'S WHAT YOU NEED

- ❏ 6 sheets of construction paper: 3 colors of construction paper, 2 sheets of each color
- ❏ Scissors

HERE'S WHAT YOU DO

1 Take one piece of each color and have your little scientist cut a whole bunch of squares, so that in the end you have, say, one blue sheet and a whole mess of blue squares. The squares don't have to be perfect—maybe 2" by 2"—and the amount doesn't matter all that much. You just need enough to scatter over the uncut construction paper. Repeat this process with each color until you have three sheets of uncut paper and a mess of colored squares. *(See Fig. 1.)*

2 Now the really fun part: choose one piece of construction paper and scatter a bunch of the different colored squares on the sheet, while your little scientist closes her eyes. Notice how some of the squares blend in with the sheet of paper? Perfect.

3 Now tell your little scientist to open her eyes and grab as many colored squares as she can in 5 seconds. Keep count and then stop the grabbing when you reach 5 seconds. Now, count how many colored squares your lab partner picked up. Were they the same color as the paper background you chose, or was it easier to perceive and grab colored squares that were different from the background? This, my friends, is camouflage in action. Try different backgrounds and different shapes of cutouts to see what the eye is better at picking out.

Fig. 1

BANANA BALLOON

I admit it: We got this experiment out of the way as fast as humanly possible. We are not a banana family. And that's putting it politely. If we see a banana walking down the street, we usually cross to the other side. They're gross and smelly and mushy enough as is, but once they become animate . . . we sort of freak out.

We may hate bananas, but they are absolutely perfect for science experiments. And sometimes in science, you just have to button up your personal fears and work with things you might ordinarily shy away from.

Bananas *are* a pretty good science tool after all. You can carry them anywhere, they're easy to open, and they contain a lot of gas. And bugs. Teeny, tiny, miniature bugs—just like the ones crawling all over *you*! (More on that below.)

The gases and bugs inside a banana are a perfect combination for an experiment about decomposition, or the breakdown of living material that will eventually happen to us all. Did you know you can blow up a balloon simply using the gases from banana decomposition? That's the gist of this experiment, so let's get to work. (And get it out of the way.)

HERE'S WHY IT WORKS

In this experiment, we'll talk a lot about *decomposition*—the breaking down of matter. You've probably seen bananas decay. You bring them home from the store and they're yellow and pretty and firm. But a week or so later, they're a brown, splotchy mess.

Well, that's called decomposition.

When you put a banana in a bottle and seal the bottle with a balloon, something fascinating begins to happen. Bacteria—tiny organisms that live in and on everything, including you, begin to munch away at the fruit. The organisms multiply and multiply, gobbling up the banana. During the process, they give off a waste called ethylene gas.

That gas will seek places to expand and will eventually blow up the balloon. (Yes, it takes some time, so be patient.) While you're waiting for the balloon to expand, use your measuring tape to mark the growth of the balloon and be sure to watch what happens during the experiment—to the balloon *and* to the fruit goop in the bottom of the bottle. (If you're a fan of bananas, you probably won't mind the brown, mushy soup at the bottom of the bottle. If, like us, you're *not* a banana fan, then this is your warning: Maybe try a different experiment instead. Or make like a real scientist and dig in.)

HERE'S WHAT YOU NEED

- ❑ A banana
- ❑ A balloon
- ❑ A bottle
- ❑ Measuring tape or string (Optional. You can just keep an eye on it.)

HERE'S WHAT YOU DO

1 This one is remarkably simple: Have your lab partner mash up a banana in the bottom of your bottle and then slip a balloon around the top of the bottle. *(See Fig. 1.)* If your lab partner wants to mash the banana before you put it in the bottle, or wants to mash it after you put it in the bottle, it doesn't matter. Whatever your lab partner wants is fine. Let him take the lead. Seriously, the prep time on this takes all of 2 minutes and it's something any tiny mad scientist can do completely on his own.

2 Then . . . you wait for the balloon to expand. *(See Fig. 2.)* You wait a *long* time. Between a few days to a week or more. So be prepared to use the measuring tape or string to turn this into a long-term project that puts observational skills and patience to the test. It's a fabulous starter project for budding scientists, as you use the tools to measure balloon growth day after day. Have your lab partner keep a log if you'd like.

WANT MORE?

If you'd like, try different fruits or vegetables with other bottles and balloons at the same time. All rotting fruit you see gives off ethylene gas, thanks to the bacteria eating away at them. Because each fruit and vegetable gives off different amounts of gas during decomposition, the effects on the balloon should be different. See which fruits or veggies cause the balloons to expand the most!

Fig. 1

Fig. 2

I'll be honest. Physics scared me in school. At first blush, it seemed like a difficult, high field of mathematics and calculations of the unseen forces that interact with us. Then I got to know it and realized just how inspiring it was to see physics at work in everyday life. Balance. Weight distribution. Motion. Give. Pressure. Energy. Gravity. We're constantly interacting with the forces of physics. Every day. Every *second*.

In the experiments in this chapter, you and your little lab partner will explore the theories that keep us moving (Fast Corners), that keep things balanced (Balancing Act), that fool our eyes (Colors of Light), or give us the ability to launch a rocket and reach the moon (Straw Balloon Rocket Blasters)—not to mentioning getting back home again. With household supplies that you can find in your junk drawer, you'll be able to perform easy experiments that study the theories of physics that explain the world around us and how we, as humans, experience it. And, as if that wasn't good enough, these experiments will let your little scientist get her hands dirty and have fun too.

STRAW BALLOON ROCKET BLASTERS

Warning: This is one of those science experiments that will have you on repeat for the rest of the day, so plan ahead.

Yes, you'll learn about rocket propulsion. Yes, you'll learn about Newton's Third Law of Motion (every action has an equal and opposite reaction). But mostly, you are going to have so much fun that you'll probably be doing this over and over and over again. Basically, you're going to strap a straw to a balloon and watch as it propels along a track of wire or thread, but again, be warned, this is addictive and soon enough your little mad scientist will be asking to launch rockets to the roof and beyond.

Science, my friend, rocks.

WHY IT WORKS

Here's the deal: You're going to run a long string through a straw and then tie one end of the string to a high place—a tree, a bookshelf, doesn't matter. Then you'll tape the straw to a party balloon you simply blow up. Hold the other end of the string with one hand while you let the balloon go. Watch as the balloon climbs up the string—a mini-rocket in the works.

But what makes the balloon rocket away? It's just filled with your hot air, right?

Exactly.

This is Sir Isaac Newton's Third Law of Motion in play. (He was the guy who figured out that gravity is the force that keeps us tethered to Earth.)

The law goes like this: For every action (the air escaping the balloon and pushing against the air outside), there is an opposite reaction (the balloon moving forward). Real rockets work in much the same manner, only they use fuel and enormous engines to create the thrust, or downward force, needed for the upward force of the rocket. Plus, instead of a string and a straw to keep them flying straight, they've got wings and stabilizers. (Which would also be a phenomenal science experiment: to see if you can craft wings or rocket-like devices to make the balloon go straight without the string and straw set-up—a very fun rainy-day science project.)

 DID YOU KNOW?

If you're like us, any time you bring out balloons, it's only a matter of time before your little scientist starts wondering just how many helium-filled balloons it would take to make a person fly. Well, to make an 85-pound person float, it would take a helium balloon at least 13 feet in diameter. That's a lot of helium . . .

HERE'S WHAT YOU NEED

- ❏ A balloon
- ❏ A long string
- ❏ Scissors
- ❏ A straw
- ❏ Tape

HERE'S WHAT YOU DO

1 For starters, have your lab partner blow up a balloon and then let it go. What happens? It goes crazy, right? Right. Now let's try to focus that air pressure to make a cool DIY rocket.

2 Cut a long piece of string and tie or tape one end far away. It doesn't matter how far or how high you go. Basically, the string will serve as a "track" for your rocket, so set your track anywhere you'd like. We've taped ours to the kitchen table and then stood back, and we've taped one to the top of the stair banister. Go outside and find a good tree branch. The ceiling. A high shelf. You name it.

3 Next, run the other end of your string through the straw. Give yourself a few inches of string to hold on to.

4 Then, inflate the balloon and hold it tight—don't tie it. Have your little scientist then tape the straw to your balloon, or you tape it while your lab partner holds the balloon. There's no right way. But for this step, it definitely helps to have a partner. Taping the straw and holding the balloon is tricky . . .

5 When you're ready, let the balloon go, and watch as the balloon then climbs up the string track, rocketing away.

FLOATING WATER

Can you make water float? I bet you can.

No, you don't need to be a wizard or a witch. You don't need to cast a spell. There's nothing magic about it at all, in fact.

You can make water float using good, ol' fashioned, awesome science. The "trick" to this experiment is air pressure.

In this experiment, you're basically going to pour water in a glass, put a slip of paper over the glass, turn the glass upside down, and watch as air pressure keeps the paper in place—and the water "floating" in the glass.

If you want to get really deep into this experiment, you can try to increase the amount of water each time. Or keep the same amount of water and decrease the size of your paper. Or go ahead and get a bucket and some cardboard and see if *that* will work. Fun times on a summer day for sure.

HERE'S WHY IT WORKS

Air abounds. It's everywhere. We breathe it, walk through it, jump against it. But air is almost constantly at war with, well, everything—even itself. We experience weather and wind because different stacks of air press against each other, pushing the layers in different directions.

That gentle breeze? That hurricane gust? That's air at war with itself.

Where there's air, there's air pressure. It's what makes airplanes fly and keeps your car tires rolling nimbly down the road.

Or, in this case, it's what makes water float.

In the first part of the experiment your lab partner will simply pour water out of a glass and watch gravity pull the water down and into the sink. The water fell. Surprise, surprise. You knew that was going to happen.

Ahh, but enter air pressure—the force that bests gravity in this experiment.

In the next part of the experiment—when you'll add a card to the top of the glass and then turn the glass upside down—the air pressure *under* the glass presses up against the card, creating a force strong enough to cancel out the effects of gravity and keep the water in the glass. Like I said, air pressure is everywhere.

 **W**ANT MORE?

The faster an air particle is moving, the lower its pressure. It basically doesn't have time to exert force. Here's a fun party trick to test it out: The next time you have some helium balloons on hand, hold two of them by the strings until both are at eye level and about 6 inches apart from each other. Now blow hard in the space *between* them. Notice how the balloons touch? By blowing, you just decreased the air pressure between them so that the air pressure around the balloons forces them together. Air. At. War.

HERE'S WHAT YOU NEED

❏ A small glass of water

❏ A sink or bathtub over which to do this experiment. Or just go outside. Or do it over the kitchen floor if you're really daring (and willing to clean). Up to you.

❏ An index card or piece of construction paper large enough to cover the opening of the glass

HERE'S WHAT YOU DO

1 Fill the glass with as much water as you'd like. No need to be precise. Now have your lab partner turn the glass upside down, or maybe over her head on a hot day. *(See Fig. 1.)* That's right. Go for it. See what happens. The water pours out, right?

2 Now let your junior mad scientist fill the glass again. No need to measure, although for this part, the more you add, the more difficult the experiment becomes. Maybe start half full and go from there.

3 Now that you have a half-filled glass of water, have your lab partner put a card or paper on top of it and press down firmly, while rotating the cup until it's upside down. Now, have your lab partner remove her hand, leaving the piece of paper in place. Did it work? Did the water remain in the glass? If this doesn't work for you right away, try a larger piece of paper, or less water and watch as the water stays in place.

Fig. 1

FAST CORNERS

In our family, I stay home to care for my own little lab partner, Emme. I'm in charge of the day-to-day routine of life: the cleaning and grocery shopping, the errands and PTA meetings, the homework help and sports practices.

I'm also in charge of the car pool.

And I *love* the car pool. It's like a moving science lab.

I remember one fun car pool ride to a horse farm out in the country. The roads were twisty and the car was filled with giggling, screaming first graders, as I took the turns at probably higher speeds than I should have. I remember thinking that if I got pulled over, I had an easy explanation.

"I wasn't *speeding*, per se. We were just conducting a science experiment. About *inertia*!"

Totally would have worked . . .

The next time you're in the car, you and your lab partner can study inertia—and you don't even need to speed. You just need to take a few corners and watch—or *feel*—what happens. It's probably the easiest science experiment you can do on the fly, for free, and yet it opens up a vast conversation about the invisible forces at work in the universe.

WHY IT WORKS

You probably already know what inertia means: It's the idea that an object at rest tends to stay at rest, while an object in motion tends to stay in motion, unless acted upon by a force. Here, we thank Sir Isaac Newton for his First Law of Motion. Think about what stops things in motion. For example, the ground will stop a falling ball. A window will stop a wayward bird. Friction on the road and air resistance will bring a car to a stop if you take your foot off the accelerator. You get the picture.

So what happens to your body when you're in a car and you turn a corner? Well, if you take a left turn it will feel like your body is being pushed toward the passenger side, right? The reality is that your inertia keeps you moving in a straight line while the road pushes the car to the left. What happens then is that the right side of the car is actually turning

into *you*—and you eventually stop moving because of the seat friction on your bottom or by slamming into the car door. Remember, your body wants to keep its forward inertia. The car wants to make the turn. This collision of forces can leave you with some bumps and bruises if you go too fast. But it can also lead to an amazing amount of giggles on a twisty country road.

With a helium balloon in the car, you'll notice something unusual. Whereas you will feel as if you're moving in the opposite direction of a turn, the balloon will go in the same direction as the turn. If the car turns left, the balloon will go left. Why? Well, the helium inside the balloon is lighter than the air around it, so the heavier air forces it out of the way. Air is basically a big, heavy bully that takes up space and then tells the balloon to get out of the way, or go in the opposite direction the air goes.

 **HERE'S WHAT YOU NEED**

- ❏ A vehicle
- ❏ A curvy road (or open space for turns. Empty parking lots are great!)
- ❏ Helium-filled balloon

HERE'S WHAT YOU DO

1 Get into the car. Buckle up. Make tight turns. If you can find a parking lot where you're all alone, let your lab partner unbuckle in the back seat and flop around like a rag doll—probably like you used to do as a kid.

2 Once your lab partner has experienced the car's inertia, put the helium balloon into play. Have her hold the balloon by its string and observe its motion while you make turns.

 DID YOU KNOW?

Sir Isaac Newton came up with his Three Laws of Motion while visiting the countryside to avoid the Great Plague of London in the 1660s. His school was shut down, so he went home and began to work on scientific theories that help us better understand the world around us today.

BALLOON TOSS

Okay, so this is a tricky one.

Not because it's hard. (Besides, hard things are fun.) Not because it's really difficult to explain. But because, if your kids are anything like mine, they can sometimes be put off by lesson dropping in the middle of something fun.

You might think: "But this is a teachable moment!"

While the kids might think: "Ugh, just throw the water balloon, okay?"

So while the setup for this is easy—you're basically tossing water balloons at each other—I've found you have to be careful about how you work in the science.

Have fun and get wet first. Then, if you sense the moment is right, you can throw in all you ever wanted to impart about the Impulse Momentum Theorem. Or you can just make this a straight-up science experiment and talk all you want. Your call. But I suggest going with the flow. And watching out for flying balloons while you're at it . . .

WHY IT WORKS

So let's say someone throws a water balloon at you. If you put up your hand and just keep it nice and firm, without any give at all, the balloon will hit you and you'll get soaked, right?

Anyone who's ever been in a water balloon war knows you have to provide a little cushioning. You catch the balloon at, say, shoulder height, but then drop your hands to your hips after initial contact in order to provide a little cushioning—and, hopefully, to avoid a pop.

This is the Impulse Momentum Theorem at play—or force acting on something to stop its momentum. The tossed balloon will always have a collision. Your hand is going to stop its movement, giving it zero momentum, or, you know, stoppage. Whether this collision happens slowly or quickly is up to you.

You see, when you provide cushioning for the balloon, you increase the length of the collision between the balloon and your hand, which basically stops the balloon's momentum at a slower pace and, hopefully, keeps you from getting wet.

Think about running barefoot. Your feet have nothing to cushion them on the pavement and it can hurt. Then you put on sneakers and suddenly have cushioning. The cushioning increases the time of the collision between feet and surface, stopping the momentum a little more slowly and making it easier to run.

Have you ever tried jumping off something and not bending your knees to lessen the impact? Of course not. That's crazy! It *hurts*. You jump, bend, sometimes even tuck and roll—all of which are trying to increase the length of the collision between your body and the ground and thus soften the landing.

Whether you're tossing a balloon and trying not to get soaked, taking a walk, or playing sports with pads, you're trying to lengthen the amount of time it takes to stop something's momentum and keep yourself safe and happy. Boom. Impulse Momentum Theorem in action.

WANT MORE?

Go find an egg and a bed sheet for another example of this experiment. Have a friend hold one end of the sheet while you hold the other. Put a raw egg in the middle and see if you can use the sheet to make the egg fly into the air. The catch, of course, is to then catch it without breaking it. With a taut sheet, you're in for a mess. But with a cushioning movement, you should be okay. Also, learn to use a washing machine.

HERE'S WHAT YOU NEED

- ❏ Balloons

- ❏ Water

- ❏ Containers to soften landing for balloons (Optional. These can be something, anything, to keep the balloons from popping, including containers, bags, bubble wrap, etc. No limits.)

HERE'S WHAT YOU DO

1 Stand, say, 5 feet apart from your lab partner and toss a water balloon. After a successful catch, you both take a step back.

2 Continue tossing the balloon back and forth to see how far you can get before the balloon breaks—or the point at which your body's cushioning of the momentum can no longer make up for the force of the momentum. In other words, until you can't avoid getting wet.

3 Classroom science experiments will have your little lab partner try to devise a system that provides enough cushion to keep the balloon—or a raw egg— intact. If you'd like, when you're tired of getting soaked, have your lab partner devise a system to keep the balloon safe when you drop it. I won't mention names, but I remember a first grader who cradled an egg in a gooey batch of homemade Play-Doh and won the class contest . . . Maybe you can wrap a balloon in bubble wrap or, say, balloons inflated with air. Let your mad scientist go wild.

MENTOS AND COKE ROCKET

This is a home science project and play-date experiment that your kids will love! At its heart, it's very simple. You add Mentos mints to a bottle of Diet Coke (this particular soda works best!) and watch what happens. Trust me, you're going to be impressed by the massive, foamy eruption that occurs.

Although this is a simple experiment using simple household foodstuffs, you'd be surprised at the level of study it has spawned in the scientific community, as researchers tried to determine whether the eruption came from a chemical reaction or a physical reaction. Even the popular show *Mythbusters* conducted Mentos and Coke experiments. So the next time you're at the store, stock up on the supplies and get ready to spend an afternoon playing mad, happy scientist in the backyard. But take note: Make sure you have a hose on hand to wash away all the excess soda and candy.

Science is fun. Household ant invasions?

Not so much . . .

HERE'S WHY IT WORKS

In short, when you combine the Mentos and cola, an eruption occurs. Instead of a chemical reaction—the event that occurs when the ingredients of the mint and soda combine—the eruption comes from a physical reaction called *nucleation*, which basically means that you've given the gas in the liquid a place to form bubbles. You see, soda is made fizzy when carbon dioxide gas is added into the water. Because water molecules like to stick very tightly together there typically isn't a place for the gas to form bubbles so the gas remains suspended in the water. But then you add the Mentos . . .

These mints are unique because their surface is covered with millions of tiny craters and holes: the perfect little places for the suspended gas to suddenly cling to and form bubbles. When you put your finger in the soda or in bubbly water, you'll see how bubbles suddenly form on your skin. *That's* nucleation in action. The same thing happens with the Mentos mints, but on a much, *much* larger scale. Plus, the mints sink, allowing gas bubbles throughout the bottle to form on the mints' downward journey. More bubbles means more gas is being released and pretty soon all those bubbles add up and voilà: *eruption*!

⇨ HERE'S WHAT YOU NEED

- ❏ Mentos mints

- ❏ Diet Coke (The bigger the bottle the better; again Diet Coke works best but feel free to experiment with other liquids.)

- ❏ Rocket materials: used paper towel or toilet paper roll, paper, scissors, tape, twine or floss, decorations as desired

HERE'S WHAT YOU DO

1 The rocket building and launching can all be done by your junior rocket scientist, with you on the side for help. Have her use a toilet paper or paper towel roll for the rocket body. Cut out some paper and roll it into a nose cone, applying it to the top of the rocket with tape. Cut out some more paper for side wings, which will act as stabilizers during liftoff. *(See Fig. 1.)*

2 Next, cut two long slits in the side of your rocket, parallel to the bottom and across from each other. Later, you should be able to slip a piece of paper between these slits. Turn the rocket over so that the nose is facing the ground and then add as many mints as you'd like through the bottom of the rocket—the more the better. *(See Fig. 2.)*

Fig. 1

Fig. 2

3. Then, add a slip of paper or card stock into the slits, so that when you turn the rocket back over, the paper provides a barrier and the mints won't fall out. Tape string to the slip of paper to act as a "launch cord."

4. Place the rocket on top of the open bottle and back away with the "launch cord" in hand. *(See Fig. 3.)* The mints will rest on the paper until you pull it out—at which point the mints will then fall into the bottle and the soda will erupt. The geyser is fast and fun and worth doing on its own just to see the amazing gush of gassy foam. Have fun!

Fig. 3

WANT MORE?

If you don't want to build a rocket, the absolute easiest way to do this experiment is to open the Diet Coke bottle and quickly slip in a handful of Mentos mints—and then stand back as fast as possible. You'll likely get soaked in sugary bliss if you don't.

In addition, the really great thing about this experiment is how many variables you can add. Will fruit Mentos work just as well as mint Mentos? How about other candies? Or materials? How about other sodas, or bubbly water? Have fun trying different sodas or bubbly liquids with different candies to see how high you can boost your rocket. This is a perfect experiment to work on your graphing skills, charting out which materials and which sodas create the biggest eruptions. Or, for a really messy experiment, pop a Mentos into your mouth but don't chew it. Take a small swig of Diet Coke and prepare to start foaming at the mouth! For an even cooler option, try adding a Mentos to a bottle, screwing the lid on quickly, shaking it up and throwing it in the air as high as you can—but watch out; that bottle could go anywhere. Fun times.

MAGNETIC FIELDS

Energy is all around us. Think about the food you eat. All those calories—a unit measurement of energy—give our bodies the fuel we need to learn, to sleep, you name it. That's just one form of energy. Cars, trucks, planes, they rely on energy as well.

But how about all those appliances and toys and gadgets you plug into a wall or a battery? Where do *they* get the energy to function? They're using electricity, of course—a source of energy that's been around as long as time but one we've only managed to harness for home use in the past couple hundred years.

In this experiment, you're going to learn a little about electricity and how to use it to turn everyday objects into magnets. With some spare wire and batteries, you can turn a simple screwdriver into a powerful magnet and have a contest to see who can pick up the most pieces of metal with it. Or you can find other household wares to experiment with to see if you can build an even *stronger* magnet. Using simple AA or C batteries, your mad scientist should be able to come up with something cool. It's like a science lab from a junk drawer—any *MacGyver* fan's favorite.

WHY IT WORKS

At the heart of everything—you, the sun, the stars, the book you're holding in your hands—are atoms, the basic building blocks for everything. Those atoms, in turn, are made up of even *smaller* building blocks like the following:

- **Protons:** They are at the center of the atom, the *nucleus*, and have a positive charge.
- **Neutrons:** The parts of the atom that don't have a charge, either positive or negative. They hang around the nucleus with the protons.
- **Electrons:** And then we have electrons, the negatively charged parts of an atom that *orbit* the nucleus, much like the earth orbits the sun. Depending on the atom, there can be many, many electrons, or just one or two, all in constant orbit around the nucleus.

So what does this have to do with electricity? Well, electricity is the *flow* of those electrons. As one atom bumps into a different atom, the electrons can jump from atom to atom—forming, you guessed it, an *electric current*. Have you ever rubbed a balloon against your hair and watched as your hair then stands on end? That's static electricity at work, as the electrons jump from the balloon to your hair.

With one part of this experiment, you'll show how electricity can confuse a compass by creating a competing magnetic force. When you place a wire on a navigational compass (a very small magnet itself) and then connect the wire to a power source—a battery—you're actually turning the wire into a magnet as well, creating great magnetic confusion for the compass.

With the other part, you'll make an even *stronger* electromagnet by harnessing the power of your wire to create more magnetic pull. By coiling the wire around a piece of metal—a screwdriver—you can increase the power of your magnet. Those coils are called *solenoids,* and a nice even field of them will create a magnet strong enough to magnetize other metals, such as paper clips. So break out your tools, some tape, some batteries, and other household goods and prepare to create a mini magnet factory.

WANT MORE?

Danish physicist and chemist Hans Christian Oersted discovered that electricity flowing through a wire creates a magnet. The key difference between a regular magnet and an electromagnet is that you can turn one of them off.

⇨ HERE'S WHAT YOU NEED

PART 1

- ❏ A navigational compass
- ❏ An insulated wire with bare wire on both ends
- ❏ A 1.5 volt battery (AA, C, D; those are all 1.5 volts)

PART 2

- ❏ Long piece of copper wire, insulated if you have it
- ❏ Screwdriver
- ❏ Tape
- ❏ A 1.5 volt battery (AA, C, D; those are all 1.5 volts)
- ❏ Paper clips

HERE'S WHAT YOU DO

1 **Part 1:** Tell your lab partner to lay the compass on a table so it points north. Now have her place the wire over the compass and connect the ends of the wire to your battery and watch as the compass needle freaks out and spins. *(See Fig. 1.)* Yup. That easy. (Although I think the trick might be actually finding a compass. Most people don't seem to have them anymore.)

Fig. 1

2 **Part 2:** Now, wrap the middle of your wire around the screwdriver 10 times, leaving enough wire on each end to connect to a battery. Now, tape one end of the wire to the negative contact on the battery—the "-" end. Hold the screwdriver by the handle while you then touch the free end of the wire to the positive side of the battery—the "+" end. Congratulations. You just MacGyvered up an electromagnet. See how many paper clips you can pick up with your new magnet. Keep count. Now try creating even *more* solenoids by coiling the wire around the screwdriver some more. Repeat the experiment. Now how many paper clips can you pick up? You should find that with more solenoids, you can pick up more clips. But keep going—try creating magnets out of other pieces of metal and see what else you can pick up.

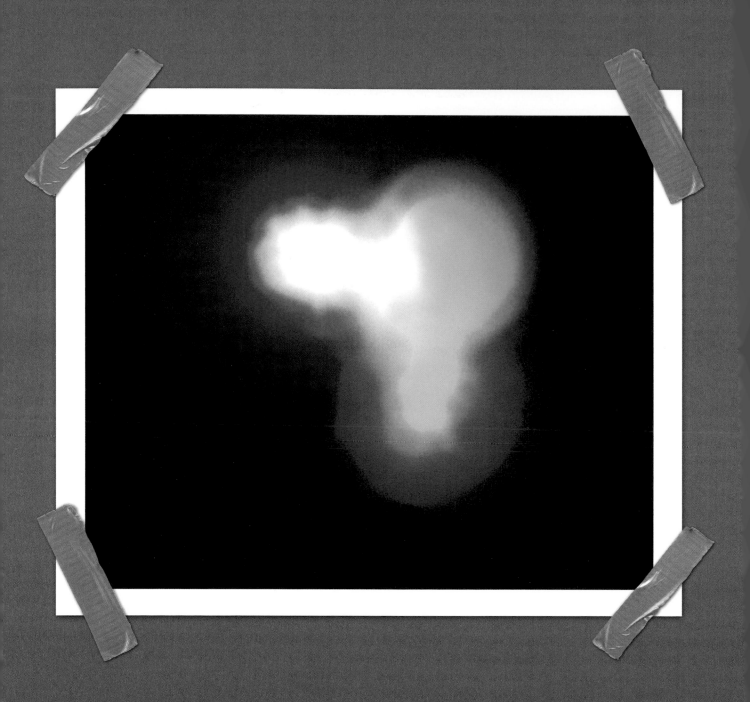

COLORS OF LIGHT

When it comes to mixing colors of paint, you probably already know the basics. Yellow and blue make green. Blue and red make purple. Been there, done that.

But what happens when you mix colors of *light*? The combinations are different than those you see when painting.

For instance, when you combine red and green paint, you get brown. But when you combine red and green light, you get yellow.

So how does *that* happen?

It all has to do with how we *perceive* light, which travels in waves to our eyes.

This fun experiment delves into primary colors and the differences between how our eyes perceive light and color. With a few household flashlights and some colored light gels—or even colored plastic wrap if you can find it—you can blend colors of light on your walls to see how they mix and sometimes even cancel each other out. It's a fun experiment if you want that little "A-ha!" lightbulb moment when it comes to explaining primary colors . . . (Okay, that was bad. Sorry about that. Carry on.)

WHY IT WORKS

When it comes to paint, there are three primary colors: yellow, magenta (a reddish hue), and cyan (a bluish hue). Go take a look at your home printer. You'll probably notice it uses inks of yellow, magenta, and cyan to produce all the other pretty colors you see when you print a photo.

When it comes to light, the primary colors are different. They are red, green, and blue. And, just like when you mix two paints together, you get a different color when you mix two lights together.

But when you mix *all* the primary colors of light together, they add up to produce . . . white! In fact, the white light that you see from, say, the sun or a light bulb is actually all the colors added together. They complement each other enough to basically cancel each other out, and we only notice different colors when that white light bounces off different surfaces to produce different colors.

⇨ HERE'S WHAT YOU NEED

❑ 3 cellophane light gels to fit over flashlights (These are sometimes sold at toy, camera, or hardware stores and sometimes come with flashlights or cameras, or you can use different colored cellophane and layer it over the flashlight.)

❑ 3 flashlights (size doesn't matter)

❑ Rubber bands to hold gels over flashlights if needed

❑ A white screen or wall

HERE'S WHAT YOU DO

1 Secure each of your colored gels over each of your flashlights and check to make sure the light is colored. *(See Fig. 1.)*

2 Let your lab partner mix two of the colored lights at a time so that the two circles of light meet and overlap enough to form a new color of light in the middle. Take note of what new color is created with each combination. Red plus green, for example, will create a yellow hue, while red plus blue creates a purple hue. When you add two primary colors, the other color you produce is called a secondary color.

3 Now here's the fun part. Carefully shine all *three* of your light colors onto the wall or screen so that they all overlap in the middle. What do you notice about the new middle color? The light should turn some shade of white. But don't worry if it's not fully white. If the cellophane light covers or gels aren't perfect, your new white light won't be either. But with even an off-white color, you should get the picture.

Fig. 1

 DID YOU KNOW?

Primary colors are the set of colors from which all other colors can be produced. There are secondary colors below them, and then tertiary colors below those.

EGG IN A BOTTLE

Stop me if you've heard this one before, but this is one of those experiments that can help you win a bet.

Simply place a peeled, hard-boiled egg atop a bottle and then bet a friend that you can push the egg into the bottle without touching it.

How, you ask?

Air pressure. That's how. Air pressure is everywhere—invisibly pushing against everything in its way, even you. To get the egg into the bottle, all you need are a few household items, a little fire or heat, and some fast hands. Or, you can crank this experiment up a few notches and use it to turn snack time into a deliciously fun cooking activity.

HERE'S WHY IT WORKS

Air pressure is like Harry Houdini, the famous escape artist; it's always looking for a way to escape, flowing from high-pressure areas to low-pressure areas. In high pressure areas, the air molecules are crammed together, while in low pressure areas they have more room to flow. When it comes to weather, low pressure areas in the atmosphere are generally warmer and associated with big storms and clouds—think tropical. High pressure areas generally happen in colder areas—think polar. When it comes to air pressure with everyday items, think about the times you've had a flat tire. The high air pressure in the tire escapes through a tiny hole into the lower-pressure atmosphere. It found an escape route. Your kids have probably seen the same thing with an open balloon that's gone pfffffttttzzzzziiinnggg away.

So what does this have to do with the egg? Well, when you light a fire and then rest the egg on top of the bottle, you're basically putting the egg between two opposing layers of air: the high pressure outside the bottle and the low pressure inside the bottle. The air outside wants to get inside, so it presses the egg out of its way to make this possible. (Air pressure is sort of a bully, to tell you the truth.)

But why is the inside of the bottle a low-pressure area? It doesn't start off that way. When you light the fire inside the bottle, you burn off all the oxygen inside and you *create* a low-pressure environment. Once this happens,

the air outside the bottle, which is always pushing and shoving against itself looking for a low-pressure area to take over, exerts so much pressure on the egg that it forces the egg into the bottle. It all happens pretty quickly, so keep a close eye on the egg.

 DID YOU KNOW?

So wait a minute, you say. What do you do with the egg in the bottle *now*? You can get it back out again just by using your *own* air pressure. Tilt the bottle until the egg is at the opening. Now put your lips around the bottle and blow. Hard. By blowing, you're creating a high-pressure environment in the bottle again. Blow hard enough and suddenly the *inside* of the bottle has a higher pressure than the *outside* and voilà, the egg comes sliding out into your mouth.

If you don't want an egg covered in burnt paper in your mouth, try lighting something like cinnamon or some other spice or herb on fire, instead of using matches and paper. That way, you're still eliminating the oxygen inside the bottle, but when you blow on the bottle to get the egg back, you suddenly have a perfectly scented egg, instead of something that is covered in burned paper. Science can be delicious, yo. Enjoy!

HERE'S WHAT YOU NEED

- ❏ A wide-mouth bottle—something that is just wide enough to squeeze a peeled egg through but not so wide that it will simply fall through. We use a glass milk bottle.

- ❏ A peeled, hard-boiled egg

- ❏ Matches or handheld torch

- ❏ A small square of paper

HERE'S WHAT YOU DO

1 First, put your egg on the bottle. It just sits there, right? You shouldn't be able to easily push it in.

2 Now, remove the egg from the bottle opening and have your lab partner get your flammables ready. Have her light your matches or torch *(see Fig. 1)* and paper and drop them in the bottle, placing the egg quickly back on top to seal the bottle. Don't fret if you don't see flames, because even a bit of smoldering paper will do the trick. *(See Fig. 2.)*

3 Watch closely. This happens fast. Before you know it, the egg will get sucked into the bottle with a nice little *pop*!

Fig. 1

Fig. 2

BALANCING ACT

Seesaws, or teeter-totters, have changed a lot over the years. It used to be that they were made of wood and gave your rear end splinters from time to time. But those splinters weren't nearly as dangerous as what happened when a friend jumped on the other side when you weren't quite ready, sometimes catching you unprepared for the sudden lift on your end.

You know what always happened next: *thunk*! A swift uppercut to the jaw.

The typical playground seesaw may have changed enough over the years to become relatively safe—you have to work pretty hard to get hurt—but the physics at play remain the same. In this experiment, you'll replicate the playground seesaw with school equipment and pennies, explore the idea of balance, and tinker around with one of the simplest machines ever invented.

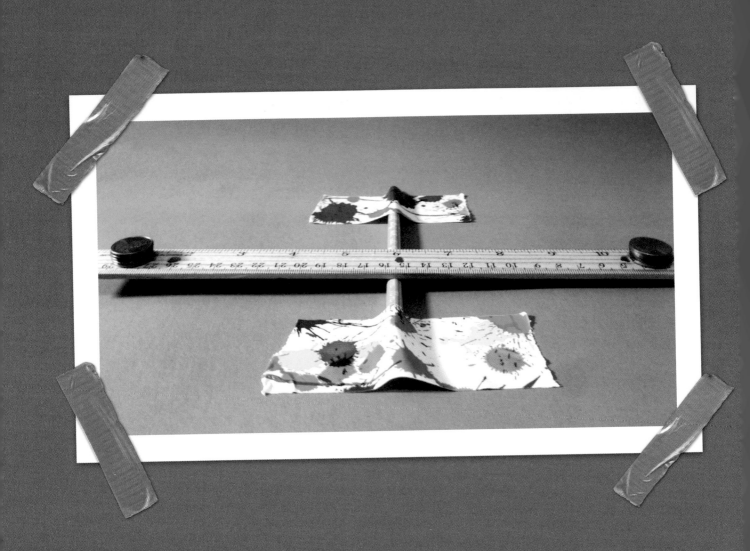

WHY IT WORKS

The short story looks like this: You're going to balance a regular 12" ruler atop a pencil, trying to find a balance point so that both ends of the ruler hover above your table or desk. Then you're going to add weights—pennies—on each end to see if you can still keep it balanced.

The pencil turns the ruler into a lever—one of the simplest mechanical devices. You've probably used levers before: a screwdriver to pry something loose, a crowbar, a toothpick. You get the idea.

But a lever alone won't do the trick. It needs a pivot point, something to help give the lever . . . leverage. In this case, the pencil acts as a *fulcrum*, or the pivot point. To balance the middle of the ruler on the pencil, you need the same kind of force on each end of the pencil. The force is gravity, pushing down on both ends of your ruler.

When you start to balance pennies on the ruler, there's a catch. The farther away you place the pennies from the fulcrum, the more their gravity counts in balancing them. Three pennies located 4 inches from the fulcrum will balance out 6 pennies located 2 inches from the fulcrum, for instance. Think of this equation on how it balances it out: 3 pennies × 4 inches on one end = 12. On the other end, it's 6 pennies × 2 inches = 12. *Balance!*

When it comes to balancing acts, there are three classes of levers. You've probably used all of them in everyday life:

- The pencil-ruler balance, with the fulcrum in the middle, is called a Class 1 lever. Think seesaw.
- Class 2 looks a lot like a wheelbarrow, with the fulcrum (the wheels) on one end of the lever and the force (you lifting the handles) on the other end.
- A Class 3 lever in action is a lot like a pair of tweezers—the fulcrum on one end and the effort applied to the middle, where you pinch to make the ends close.

With some school supplies and spare change, your little lab partner can tinker around with levers and find the perfect balance points.

HERE'S WHAT YOU NEED

- ❏ A pencil

- ❏ A ruler

- ❏ Tape, to hold your pencil in place (optional)

- ❏ 10 pennies (Find ones minted after 1982, so that they all have the same metals and weight.)

HERE'S WHAT YOU DO

1 Start with your pencil and a hard, level surface—a desk or table, for instance.

2 Now balance your ruler over the pencil at the 6-inch mark. Try to get both ends of your ruler to hover above the desk. If needed, you can tape the pencil down to keep it from moving around.

3 Now place 5 pennies on one end of the ruler and observe what happens.

4 Take 5 *more* pennies and try to find the location on the *other* end of the ruler to bring the lever into balance again—or the equilibrium of the miniature seesaw.

5 Clear the ruler and try again. This time, put 6 pennies at the 2-inch mark of the ruler.

6 Now try to find the location on the *other* end of the ruler in which 3 pennies will balance out the other 6. Tinker around until you find it. In our lab, it took some time and some shifting around of the pennies to get it just right, but there's always a high-five moment when we do. You can do it.

WANT MORE?

The next time you're at the playground and see a seesaw, grab your lab partner and try to find the point where you need to sit on your end of the lever to balance out the equipment. Then, if the seesaw is old-school enough, try not to get your teeth knocked out when you both decide to try something else and get off the toy . . .

You've heard of Goldilocks I'm sure: that adventurous little lockpick and sneak thief out to find the meal or chair or bed that is "just right."

Planet Earth has already found this zone—this "just right" positioning in the universe that allows it to bear life. We're not too close to the sun, not too far away either. Scientists describe planet positions like ours as being in the Goldilocks Zone, and it's quite possible there are other planets out there revolving around their own stars in this perfect habitable zone that allows life to flourish.

But until we find them, Earth, it seems, is unique.

We've got water, air, plants, animals—abundant life to keep us all going. Of course, one of the biggest threats to all this comes from one of the animals that has evolved on the planet and done some amazing good but also some amazing damage: *us*.

In the experiments in this chapter, you and your little lab partner will explore the wonders of our natural habitat (Land Warmer). We'll also take a look at the damage humans have done to the planet (Acid Rain) and what we can do to help reverse that damage so that our planet—and we—can continue to thrive.

VOLCANO TIME!

If you grew up watching endless *Brady Bunch* reruns you're probably familiar with Peter Brady's volcano—a mud-spewing, steep-sided science project that sent showers of muck and sludge all over Peter's sister, Marsha, and her snooty friends. It was the coolest thing ever.

There's a good chance that this one episode alone launched our love affair with kitchen-sink volcano projects—an experiment so simple that you and your lab partner can most likely do it right now with stuff you already have in the kitchen. All you really need is vinegar, baking soda, and a bottle to mix them in, but it is much cooler to use good ol' fashioned backyard dirt to construct a volcano model around the bottle first and *then* conduct the experiment.

Either way you do it, this is a science experiment with serious thrills. But it also expertly mimics what happens under the earth's crust to create volcanic eruptions.

HERE'S WHY IT WORKS

When the solid baking soda (sodium bicarbonate—a base) mixes with the liquid vinegar (acetic acid—a weak acid), a chemical reaction occurs and forms a gas (carbon dioxide). All those bubbles and foam? They're evidence of gas, and as the gas expands, it looks for an escape route for all that built-up pressure. So the foam and bubbles rise until they flood out of your bottle's opening.

Pretty much the same exact thing is happening under the earth right now.

The earth's crust is made up of many sections of superthick shell—65-plus *miles* thick!—called tectonic plates that are always moving, very slowly, over the much, much hotter inner earth. Most of the world's volcanoes are found where two or more of these tectonic plates meet one another. Sometimes those plates shift and sometimes they collide, forming escape routes in the earth's crust for molten rocks and gas, called magma. Much like the carbon dioxide in your baking soda–vinegar experiment seeks the quickest escape route to relieve pressure, the gases in the underground magma do the same thing before erupting out of a volcano.

Not all eruptions are alike, however. Sometimes the gases in the magma are easily released from the earth's crust and the result

is a slow, oozing spread of superhot lava. But sometimes the gases stay trapped beneath cooled magma and rock building up pressure until they erupt in violent explosions that can send ash and boulders flying up to 20 miles high. In fact, airplane pilots keep track of volcano activity around the earth, just to be sure they don't fly into clouds of dangerous ash.

HERE'S WHAT YOU NEED

- ❏ Baking soda
- ❏ Vinegar
- ❏ A bottle (a good vase with a wide bottom and slender top also works well, but use whatever you can find)
- ❏ Red food dye
- ❏ String
- ❏ Toilet paper

HERE'S WHAT YOU DO

1 First add the vinegar to your bottle and dye it red with food coloring. Then, rip out a few sheets of toilet paper and make a pouch for the baking soda. Use your string to tie the pouch and then insert the pouch into your bottle, using the bottle cap to hold the other end of the string so that the pouch dangles above your "lava." *(See Fig. 1.)*

2 If you're feeling super science-y/crafty, let your lab partner shape a volcano model out of backyard mud and dirt around the bottle. Note: you don't have to do this, but go big or go home, right?

3 When your volcano model is ready, lift the cap and watch the pouch drop into the lava. It will foam up slowly, mimicking the slow buildup of earth's gases, until the vinegar fully soaks the tissue paper. Then, the fun really begins, as the foam begins to climb the bottle, looking for an escape route. Just stand back, and watch the foam erupt. It's really that easy!

Fig. 1

 WANT MORE?

There are many, many ways to perform this experiment, so don't be afraid to get creative. Try mixing the vinegar and baking soda in a bottle, and then quickly place a balloon over the bottle opening. While this doesn't create a lava explosion, the gases will inflate the balloon. Pretty cool, right? Or, put baking soda in small snack-size zip bag and seal tightly with a bit of air in the bag. Place the baking soda bag in a larger zip bag that is filled with vinegar and seal that bag tightly, with as little air as possible in the big bag. Now use your fist to smash the tiny baking soda bag and stand back. You just made a sandwich bag bomb, using the same chemical reactions as your volcano.

ACID RAIN

When we burn coal for power or gasoline for cars, we put pollutants in the air. These pollutants react with the water in the air (rainwater) and form an acid.

This is acid rain. You may have heard about it.

It looks, tastes, and *feels* just like regular rain. But it's full of so many fine particles of pollutants—sulfur dioxide and nitrogen dioxide—that it causes heart and lung damage to humans when the particles are inhaled. It also wreaks havoc on wildlife and our ecosystem by altering the acidity of lakes, rivers, and forests.

It definitely pays for all of us to consider alternate forms of energy creation, such as solar or wind power.

WHY IT WORKS

In this simple experiment using common pool supply materials, you and your lab partner will be able to see the impacts of carbon dioxide on water in a matter of seconds. You'll need a chemical called phenol red. This is the same stuff you use to measure the cleanliness of your pool to determine whether you need to add more chlorine. You can find this at a pool supply store, or perhaps your neighbor or community pool might loan you a few drops.

For this experiment, you'll add phenol red to a glass of water and then use a straw to blow into the water. You'll notice the water turns from red to clear, as you add carbon dioxide—the gases you exhale—and forms a weak acid in the liquid.

The same thing happens on a much larger scale in the atmosphere, as pollutants meet and mix with moisture in the air and create an acid that is then rained down upon the earth. This is not good for the future health of our planet.

Maybe your little scientist can think of more alternative forms of energy that are strong enough to keep us moving but healthy enough to keep us around longer?

 **HERE'S WHAT YOU NEED**

- ❏ Phenol red—about 20 drops
- ❏ A glass
- ❏ A straw
- ❏ Some water

HERE'S WHAT YOU DO

1 Fill your glass about three-quarters full.

2 Add about 20 drops of phenol red, or enough to turn the liquid light red. You might need to tinker with the amount over the course of a few experiments until you get it just right.

3 Now slip the straw in the water and let your lab partner blow hard enough so that bubbles form but don't tumble out. Note: Don't drink or touch it, as it can be an irritant.

4 Keep an eye on the color. It should start changing to a much lighter shade of red as you continue to pump in more carbon dioxide. Repeat the blowing until the water clears up.

 DID YOU KNOW?

The United States Environmental Protection Agency says it is critical to reduce acid rain, or *acid deposition* as it is called. By conserving power at home, you can help reduce the amount of coal burned at the power plant and thus the pollutants released into the air. But that's just one small step. We need millions of people, around the world, to do the same thing.

LAND WARMER

Ever walk on the beach on a superhot day? Notice how your feet practically blister in the sand and then cool off in the water?

Both surfaces—the land and the water—are under the same heat from the sun: solar radiation. But the land feels considerably hotter, right?

This experiment shows why land heats faster than water, and it is the perfect way to work on you and your lab partner's graphing skills.

Plus, it's something your lab partner can do pretty much all on her own with a few household supplies and a trip to the backyard for some dirt. Those are always the best experiments, in my opinion: those cool things you can MacGyver up with stuff you already have.

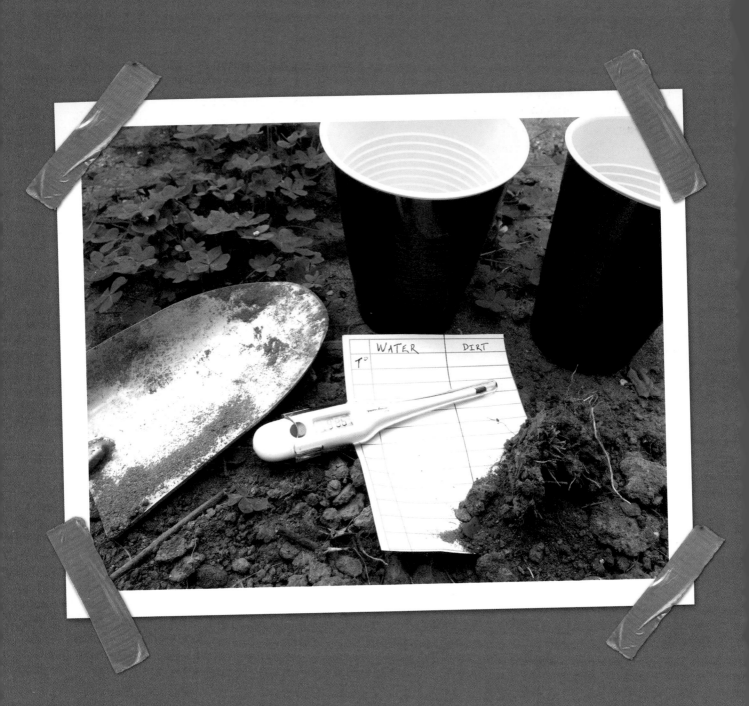

WHY IT WORKS

In this experiment, you're going to freeze two cups—one filled with water and the other with soil—for 10 minutes and then place each cup in the sun to see which warms up faster. Seriously. That's it. Easy peasy.

You'll notice that the soil heats up faster than the water, because water requires significantly more energy to heat than land. In other words, it has a higher heat capacity—or how much heat is required to raise 1 kilogram of something (water, air, soil, metal) by 1°C.

Think of the sandy beach again. The sand burns your feet while even a thin trickle of beach water is enough to cool both them and the sand near the shore. It takes *more* energy to raise the temperature of water than it does to raise the temperature of land. So until the sun becomes a giant, Earth-consuming ball of heat *billions* of years from now, you'll always have a cool place to go dip your feet after a run on hot sand . . .

HERE'S WHAT YOU NEED

❏ 2 small cups

❏ Water

❏ Dirt

❏ Thermometer (2 if you have them, but no big deal if you don't)

❏ Paper and pencil for graph (optional)

HERE'S WHAT YOU DO

1 Have your little lab partner fill both of the cups—one with water and one with dirt. Just go out to the back or front yard and fill 'em up.

2 Now place both of your cups in the freezer and wait for 10 minutes.

3 Now comes the fun part. Take out each cup and place in direct sunlight. Record the initial temperature in each cup as you do this. Just dip a thermometer in each cup and mark the reading. Most people seem to have one digital thermometer nowadays, and that's cool. Just clean it off before swapping cups.

4 After 5 minutes, take the temperature of the cups again. What'd you find? Keep taking the temperature every 5 minutes and graph your results for, say, 30 minutes. Again, you'll notice that the cup of dirt heats up much faster than the cup of water, showing the heat capacity of each medium. If you'd like, have your lab partner graph out the temperature changes for each time period. She can set up a graph any way she likes, but for starters, it might be helpful to put land in one column, water in another, and then mark temperature changes under each column, with time increments on the side.

 DID YOU KNOW?

There are many different ways, especially in science, to measure temperature and heat. You're probably already familiar with Fahrenheit, which is the system used in the United States to measure temperature. Its freezing point is 32°F and its boiling point is 212°F. All other parts of the world use Celsius, which has a freezing point of 0°C and a boiling point of 100°C. In America, your normal body temperature is 98.6°F. In the rest of the world, it's 37°C.

THE SPACE OF AIR

If your lab partner is more than two seconds old, you've probably already encountered this question: Why is the sky blue?

And, of course, you know the answer: Light travels in waves. Sunlight is white—made up of all the colors of the rainbow. Once sunlight hits our atmosphere, the white light bounces off molecules in the air and scatters. The smaller, shorter blue waves are scattered more than other colors, and that's why we perceive the sky as blue. Unless, of course, it's red. Or orange. Or purple. (Again, all to do with how white light is scattered when it hits the atmosphere and any pollution in it.)

As your junior mad scientists get older, the questions become more and more challenging. One of my favorites is: How do you know that air is . . . well, *there*?

You can't see it. You can't smell it. You can only feel it when the wind blows. So does it take up space at all?

Let's take a look.

WHY IT WORKS

This simple experiment, using a spare bottle and a balloon, will help reveal how air takes up space by showing what changes happen to it under different temperatures.

When you put a balloon over a glass bottle, you capture the air inside the bottle. The balloon doesn't inflate, it just fits over the bottle and sort of dangles a little.

I know, I know: It's not much, but it's a start.

Now, when you heat up the bottle—by placing it in boiling water—you heat up the air inside the bottle and it expands. With nowhere else to go, the air rises into the balloon and the balloon inflates. It's not magic by any means. It's all about expanding air, showing you that although you can't *see* any air in the bottle, it is indeed there.

Now when you remove the bottle from the boiling water and put it in an ice bath, the air compresses and the balloon deflates. Not only that, it gets sucked into the bottle. (Try jiggling the balloon after a few minutes to help it along if you need to. Sometimes there's a happy little *pop*! as the balloon is sucked in.) Then when you just let the bottle return to normal room temperature, the balloon should return to its starting position, shape, and size. In this part of the experiment, the air inside the bottle is growing and shrinking, inflating and deflating the balloon—giving you a firsthand glimpse at the properties of the space all around us.

Now, when you add the funnel to the experiment, you have another way to see how air takes up space and can't be easily squeezed. When you seal the funnel to the bottle with Play-Doh or modeling clay, you're not giving the air inside the bottle an escape route. When you pour the water in, the air can't go anywhere, so it just stays in the bottle. But the water isn't heavy enough to push its way down through the air so, instead, the water remains in the funnel while the air remains in the bottle. Then, when you go ahead and insert a straw into the bottle through the funnel, you give the air an escape route and *whoosh*! Down goes the water because there's nothing left there to counteract the effects of gravity.

Cool, right? Sometimes I just want to wrap my arms around science and give it a hug for being so fun. "Hey, science, I like you."

HERE'S WHAT YOU NEED

- ❏ A small-mouth glass bottle
- ❏ Balloon
- ❏ Pot of boiling water
- ❏ Pot of ice water
- ❏ Funnel
- ❏ Play-Doh or modeling clay (anything to create a seal between a funnel and a bottle)
- ❏ Water
- ❏ Straw

HERE'S WHAT YOU DO

1 Have your lab partner slip the balloon over the mouth of the bottle. Make sure there's a good seal. It should just sort of dangle. You're off and running!

2 Now gently place the bottle in the pot of boiling water and watch what happens to the balloon. Have your lab partner take notes or make observations as the balloon inflates. *(See Fig. 1.)*

3 Now remove the bottle from the water. It shouldn't be all that hot. It only takes a few moments for the balloon to inflate.

4 At this point, it's time for the soak. Put the bottle and balloon into the ice bath and note your observations. (Pay attention to where the balloon winds up. It should shrink and pop inside. Feel free to help it along with a tiny tug or shift, if you or your lab partner is growing antsy.) *(See Fig. 2.)*

5 When you're ready, remove the bottle and set it on the counter for 10 or 20 minutes—or until it returns to room temperature—and note what happens to the balloon.

6 And now comes the funnel part. Remove the balloon from the bottle and replace it with the funnel. Now pour some water in. What happens? It just pours in, right?

7 Now take your Play-Doh or modeling clay and seal the funnel to the bottle. This part is surprisingly tricky. You have to make sure there's an absolute airtight seal or it won't work. Hence, the clay. (We tried duct tape, balloons, shrink-wrap, you name it, and always seemed to have leaks.) Be creative. Just be sure there's a good seal between the funnel and the bottle so no air can escape where the funnel rests on the bottle.

8 Before you add water to the funnel, ask your scientist what will happen. Then add the water and see! (For the most part, the water should remain in the funnel, held up by the air pressure inside the bottle. A few dribbling drops is fine. Don't beat yourself up.)

9 Now insert your straw through the funnel and into the bottle, and watch what happens when you give all that air an escape route. Instant waterfall!

Fig. 1

Fig. 2

 DID YOU KNOW?

While humans perceive the light-scattering sky as blue, bees see the sky as ultraviolet because they have different light-sensing eyes than we do. So here's a mind-twister for you: If humans see the sky as *blue*, and bees see it as *ultraviolet*—which color is it *really?*

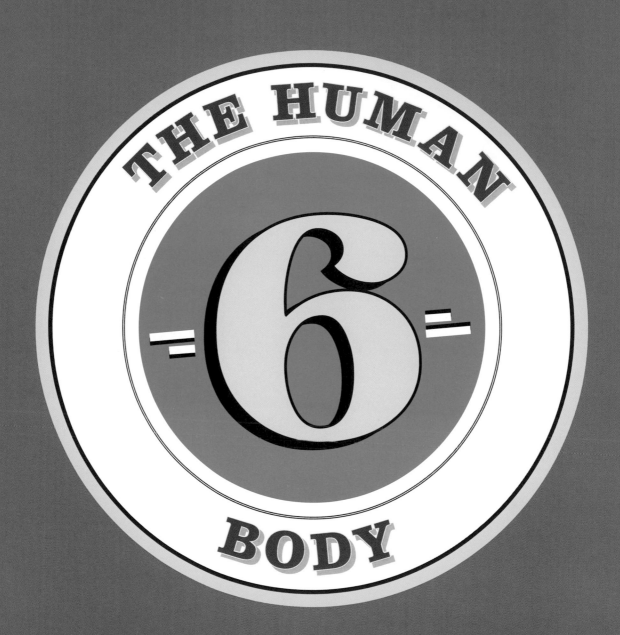

THE HUMAN

6

BODY

Your body is the coolest science lab going. No doubt about it.

For one thing, it's free. You don't have to go to the store to pick up a kit or materials. You already have everything you need for some amazingly fun science experiments.

For another, it's simply fascinating.

How does your body pump blood? How do you see? How do you balance? There are millions of questions to explore when it comes to your body—and you're probably already getting a lot of those questions from your little scientist.

In the experiments in this chapter—ranging from Blind Balance to Birdcage to Fingerprint Monsters—you'll explore your senses, the methods you use to interact with and understand the world around you. So get ready to move your body and have fun experimenting with the coolest lab you'll ever see!

HOT AND COLD

Quick quiz: What's your body's biggest organ?

Think about it for a second. Go on, I'll wait.

Okay, what'd you come up with? Your brain? Your heart? Your stomach? Nope! It's actually your *skin*.

Your skin is your largest organ, and a typical adult carries about 8 pounds of it, while kids have 6 pounds or so. Stretched out, it would cover 22 square feet. It does everything from keeping your innards and blood and bones from oozing all over the place to regulating your body's temperature. When you get hot and start to sweat? That's *perspiration*, which helps keep you cool—and also helps your body get rid of wastes.

Your skin also plays the biggest role in saving your, well, skin. When you touch something superhot and can hear your brain shouting, "Ouch!" that's your skin saying, "Stay away!" Your touch nerves are your first line of defense when it comes to keeping you safe.

But how does your skin recognize things like temperature or pain?

Let's take a quick look with a simple experiment that lets your lab partner get messy by dipping her hands into bowls filled with water of varying temperature and testing out the environment-sensing power of skin.

WHY IT WORKS

With this experiment, you'll have your little helper dip her hands into various bowls of water: one hot, one cold, one at room temperature.

You give it a try, too.

Your lab partner will dip one hand in ice water and the other in warm water, before moving *both* of them to a bowl filled with room temperature water. The hand that *was* in the ice water will suddenly feel warm, while the hand that was in the super-warm water will suddenly feel cold.

It's all about perspective. Since both of your hands grew accustomed to their own environments—either hot or cold—your touch nerves sent a series of signals to your brain that things had suddenly changed once you plunged both of your hands into the room-temperature water. The same thing happens after a shower or a bath. Have you noticed? Your body is used to the warm water. But then you hop out of the bath or turn off the shower and voilà! Your touch nerves suddenly start screaming, "Things have changed!" And then you shiver.

Why does this happen? Well, your skin is one big sensory organ, among other things. It is laced with nerve endings that connect to your nervous system, which makes its way up to your brain. When you touch something or feel something like temperature changes, the nerve endings buried in your skin send signals to your brain about the changes they have suddenly noticed.

Have you ever experienced goose bumps when you suddenly get too cold or get a chill? The scientific name for goose bumps is the *pilomotor reflex*—which basically means that teeny tiny muscles in your skin called *erector pili* start to pull on your skin hair until it stands up, creating those little bumps. But why? Just as your body releases sweat when it needs to cool down, those tiny muscles tighten when you're suddenly cold to keep warm blood trapped deep inside your skin layers. Those tiny muscles are basically working to keep you warm. You probably won't get goose bumps from this experiment, but there you go anyway—more cool information about the mysteries of your own skin. Let's explore some more.

HERE'S WHAT YOU NEED

❑ Three bowls of water: one filled with ice, one with very warm water, and the last with room-temperature water. Make sure the bowls are big enough to stick your and your lab partner's hands in. You don't have to do this experiment at the same time, but you certainly could.

❑ Hands!

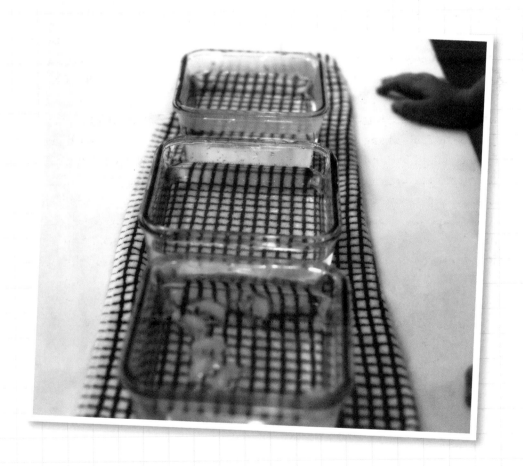

HERE'S WHAT YOU DO

1 Arrange your bowls so that the room-temperature bowl is in the middle. Now have your lab partner stick her hands in the outer two bowls—one in the super-cold bowl and the other hand in the super-warm one.

2 Keep her hands in the bowls for about 30 seconds.

3 Now have her move both hands into the center bowl, the room-temperature bowl. *(See Fig. 1.)*

4 Have her explain what she's feeling.

 DID YOU KNOW?

Your skin has three main layers: the epidermis, the inner dermis, and the subcutis. The epidermis is the outside layer and measures from 0.5 millimeters thick (your eyelids) to about 4 mm thick (your palms and hands). The inner dermis is where all the action is, packed with blood vessels, sweat glands, touch nerves, and all the oil glands that keep you from drying out. The subcutis is the base layer of fat that helps insulate you and also helps cushion you when you take a tumble. You want to know something even cooler? You probably already know that a snake sheds its skin. Well, so do you. Cells on the epidermis dry and flake off constantly. About every thirty days, you produce a completely new epidermis. Sssssssss.

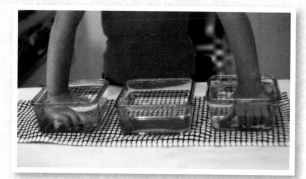

Fig. 1

BLIND BALANCE

You don't need an expensive kit or new toy for supercool science experiments. You already have the coolest, most high-tech science lab in the world.

How, you wonder? *Where*, pray tell, is this fancy, supercool science lab? Well, look in the mirror.

Your body is one of the best science experiments out there. That's right, your body. It's pretty awesome.

Ever vomited or at least felt dizzy on a roller coaster? Ever taken a road trip and felt carsick? Or went sailing and became seasick? Elevators, for some reason, get me every time.

Your body is constantly at work to keep you balanced, and sometimes signals from your eyes, ears, and muscles overwhelm your body, causing dizziness or, at the very least, imbalance.

This super-fast, fun experiment shows how balance works and questions whether you can remain balanced with your eyes closed—a query that requires no equipment at all. You're just going to stand there. It's the perfect schoolyard experiment or challenge for your little lab partner.

Fig. 1

HERE'S WHY IT WORKS

The human body is an incredible instrument that is designed to keep you upright and balanced. There are lots of tiny functions at work in your eyes, your ears, your muscles, and your nerves when you try to stay balanced. They work in harmony to sense changes and then react to them to keep you upright.

You might be wondering about your ears. How does hearing have anything to do with balance? Well, your ears do so much more than simply translate sound waves into nerve impulses that your brain then translates into the jibber-jabber of everyday life. You see, deep within your inner ear are two jelly-filled sacs that help regulate your balance. Tilt your head one way, and the jelly oozes against tiny hair sensors, which then transmit nerve impulses to your brain, basically saying, "I'm moving!" Tilt your head the other way, and the jelly oozes in the other direction and your brain gets another signal altogether. Another way to envision this system is to find a level, one of those construction tools used to determine whether surfaces are flat. See how the bubble and water inside the level move? That's like what's inside your ear. Sorta. (It's interesting to note that these sacs also contain tiny crystals as well, so the next time someone jokes, "Is your head full of rocks or something?" you can say, "Well . . . sort of!" It's also interesting to note that this usually doesn't stop the jokes.)

Things like carsickness or seasickness come from when your ears and eyes—which also send your brain signals about where you are in relation to the world and whether you're moving—simply get overwhelmed by stimuli. If you're riding in a car and reading a book or, say, can't see past the front seat, your ears tell your brain that you're moving. But your eyes tell your brain that you're standing still and reading or staring at something that is not moving, the seat in front of you for instance. Medical researchers say it's helpful to stare at the road ahead when feeling motion sickness, so your eyes and ears can both sense movement.

 ## HERE'S WHAT YOU NEED

❏ A lab partner. That's it. You guys are science all-stars.

HERE'S WHAT YOU DO

1 This is easy. Tell your lab partner to stand on both feet for 30 seconds. *(See Fig. 2.)* Easy peasy, lemon cheesy.

2 Now have her close her eyes and remain standing on both feet for 30 seconds. *(See Fig. 3.)* Still easy? Difficult? Give her a moment to write down her findings or talk about them. Some kids can stand straight up. Some will bob and weave. Some may tumble. All bodies are different, and there's no right way to do this—just have fun.

3 Okay, now it's going to get a little tricky. Have her stand on one leg for 15 seconds with her eyes open and nothing for support. *(See Fig. 4.)* Easy? Difficult?

4 Now have your partner try closing her eyes again and stand on one leg for 15 seconds with no support. Be honest. Could your partner do it? Could you? Did you crash into the couch? Take out the dog? *(See Fig. 1.)*

5 Now, try standing on one leg with your eyes closed and time yourself to see how long you and your lab partner can last without tipping or putting your other foot down.

6 Now try standing on one leg close to a wall. Keep your eyes closed but use one finger to touch the wall. This usually helps give you some balance, but again, all bodies are different, so go investigate.

7 Finally, have your lab partner stand with his legs spread apart and give him a little shove sideways. Now have him put his legs close together and then give another gentle shove from the side again. He might notice that with his legs spread, he has a lot of balance and stability, but with his legs close together . . . look out!

Fig. 2

Fig. 3

Fig. 4

 DID YOU KNOW?

For thousands of years, people have believed that eating ginger can cure motion sickness, although it seems the research is still out on that one.

BIRDCAGE

Look, you're not supposed to look at the sun. It's really bad for the eyes. But I get it. Everybody does it from time to time. You can't *help* it. There's an enormous shining star in the sky and sometimes you can't help but look.

But have you ever noticed that, after you take even a glimpse at the sun and then look away, the image remains? It's like a burning ghost version of the sun stays in your vision for many seconds.

It's called an afterimage.

And it happens all the time.

WHY IT WORKS

Your eye is an amazing, light-sensitive machine. The retina, in the back of your eye, is responsible for absorbing the light and telling your brain all about it.

Some of the light and color receptors in your retinas—called rods and cones—can become fatigued when they look at something for a little bit, even if you only look for a few seconds. Then, when you look away, the light sensors that *weren't* fatigued suddenly start to pick up the slack, and an afterimage remains. You've probably noticed a big ball of light in your vision after looking at the sun.

This experience is also at play with optical illusions. Stare at basically any color and then look quickly at a white space, and the image will likely remain in your view.

The bird-in-a-cage experiment is a fun play on this experience.

On one side of a slip of paper, you can draw a bird. On another slip, you can draw a cage. When you put the cards together and then spin them, it will appear as if the bird is actually *in* the cage—a pretty cool afterimage experiment you can do in about 10 minutes.

And if your lab partner thinks this is cool, consider all the holiday crafts you can make with this experiment. A ghost in a grave. A gold coin in a big pot. An arrow firing into a heart. There are lots and lots of afterimage craft options!

HERE'S WHAT YOU NEED

- ❏ Paper
- ❏ Scissors
- ❏ Coloring markers, pencils, crayons, you name it
- ❏ Tape
- ❏ Pencil or pen

HERE'S WHAT YOU DO

Fig. 1

1 Your lab partner should take the lead. Tell her to cut the paper into two rectangles, maybe 5 inches across and 3 inches down—about the same size as an index card. Again, the measurements don't have to be exact. Just make sure the rectangles are roughly the same size.

2 Now draw a bird in the center of one rectangle and a birdcage in the center of the other rectangle. *(See Fig. 1.)* Make sure the cage is big enough for your bird. Now, tape these rectangles together on the sides so that the images are both facing outward, but leave the top and bottom open so you can fit in a pencil in the next step.

Fig. 2

3 Insert your pencil or pen in between your rectangle sandwich and tape the sandwich to the top of the pencil. You should have enough room to hold the pencil between your hands and spin it. *(See Fig. 2.)*

4 Now . . . spin it! Experiment with speed to get the bird inside the cage and then get ready to launch into your explanation of cool afterimages.

DID YOU KNOW?

Each retina covers about 65 percent of the back of each of your eyes and has two types of vision receptors, rods and cones. There are roughly 120 million rods. They are the most sensitive vision receptors but, oddly, are not sensitive to color. The 7 million or so cones, located largely in the center of the retina, are responsible for color sensitivity.

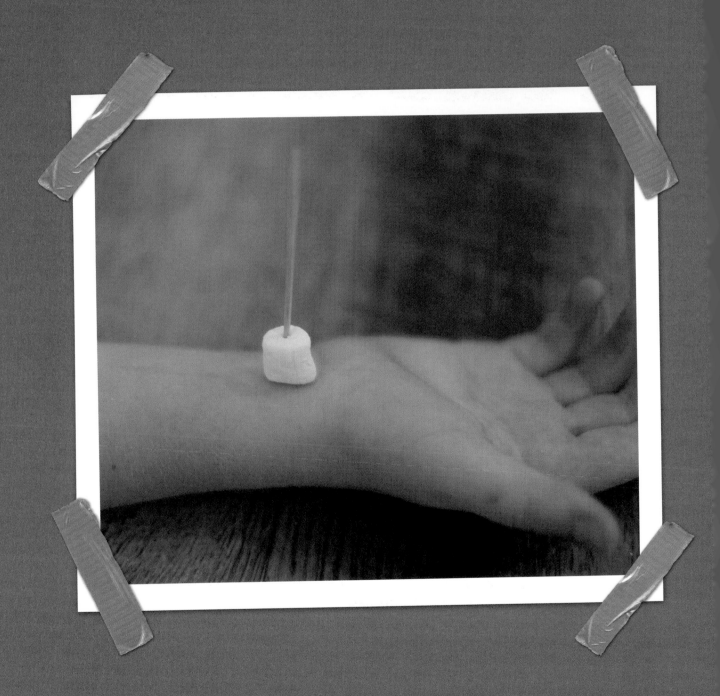

MARSHMALLOW PULSE KEEPER

Have you ever been standing or sitting perfectly still and suddenly you can feel your thumb start to . . . beat? It's almost as if you can feel blood coursing through the pad of your thumb.

Boom boom. Boom boom. Boom boom.

Well, you're not imagining things. You actually *can* feel that blood in your thumb.

Each time your heart beats, it sends rivers of oxygen-rich blood flowing through your body along a system of arteries and veins—the arteries carry blood *away* from the heart and the veins bring it back. You can feel this system at work in many parts of your body. The wrists, neck, and thumbs are great places to measure your heart rate, or the number of times your heart beats each minute.

When you're resting, your heart rate slows because the far reaches of your body don't need as much oxygen to function. But you've exercised before, right? You've felt your chest heave and your limbs tire? You've felt as if your heart was going to go thump thump thumping right out of your chest?

That's because your body needs more oxygen to function during vigorous exercise, and your heart—along with your veins and arteries and lungs—is working hard to fill the need.

Here's a fun, simple, highly jumpy experiment to see and feel your pulse rate in action and to measure just how healthy your own heart is. By sticking some marshmallows on some toothpicks, you create a pulse-measuring device you can see in action.

HERE'S WHY IT WORKS

This whole process of pumping blood throughout your body and back to your heart is done by the circulatory system, and it's what helps keep you alive.

The process is like a really cool dance inside your body each time your heart beats. In short, arteries carry oxygenated blood away from your heart to the rest of your body, coursing through a series of tubes. The tubes keep branching out and out and out until the blood flows into capillaries—even thinner tubes that allow the oxygen-rich blood to leak out and fill up each cell in your body with the air and energy needed to work. In turn, the cells feed wastes back into the capillaries and then your veins take over. The veins carry your "used" blood back to your heart and lungs, where the whole process begins again.

Take a look at the underside of your wrist. You can see the blue veins beneath the skin. Cool huh? All told, this system of blood vessels is 60,000 miles long—or long enough to stretch around the world. Twice.

 # HERE'S WHAT YOU NEED

❏ A toothpick

❏ A small marshmallow

❏ A stopwatch

❏ A pulse! (Usually found in nonvampires. If you happen to be a vampire or, say, a zombie, you can just go ahead to the next experiment.)

HERE'S WHAT YOU DO

1 Your lab partner can do all this on her own probably. Have her take a marshmallow and impale it on a toothpick. *(See Fig. 1.)*

2 Before beginning this next step, make sure your lab partner has chilled out for a few minutes beforehand. Just take a seat, relax. Maybe read a book together. This is key in finding a resting pulse rate.

3 Now tell her to place her left hand on a table, palm up. Have her place the marshmallow on her wrist—right where you see all those blue veins under the skin. The toothpick should be sticking straight up, making it easy to see your pulse in action. *(See Fig. 2.)*

Fig. 1

Fig. 2

4 Keep still and . . . watch. When you see the marshmallow and toothpick moving, start to count the number of "bumps" it makes in 60 seconds, using your watch. This is your resting pulse rate. Each bump is your heartbeat in action, pushing blood to your hands.

5 Now comes the fun part. Have your lab partner try exercising as hard as she can for 1 minute. Have her do jumping jacks. Run around the block. Run in place. Do 10 burpees. You get the picture. Exercise for 1 minute. It doesn't matter what she does, she just needs to do it.

6 Now, repeat the experiment and observe the marshmallow-toothpick. What's happening to it? Break out your stopwatch and count the number of beats again. You may find that it's quite difficult to observe the marshmallow in action after vigorous exercise. And that's fine. You can feel your pulse at your wrist, on your neck just below your jawline, or even on your thumbs. Compare your resting pulse rate to your active pulse rate. Every body is different, but your lab partner should definitely see more pulse "bumps" this time around.

 DID YOU KNOW?

When checking your pulse, you don't always have to set your clock at a full minute to determine how many times your heart beats per minute. Try instead counting how many beats you get in 30 seconds, and then multiply that number by 2. Or, count for 15 seconds and multiply by 4. Or, count for 10 seconds and then multiply by 6.

But however you decide to do it, the American Heart Association has a nifty way of finding what your maximum heart rate should be—that is, the number of beats per minute at maximum exercise. Take the number 220 and subtract your age. So if your little scientist is, say, 10, her maximum heart rate should be 210 (220 - 10 = 210!). Each person is different and each body's need for oxygen is different so don't freak out if you or your lab partner's heart rates are different from the standard. Just talk to your doctor if you're concerned. That's why she's there!

FINGERPRINT MONSTERS

You may be familiar with this scene in a movie or TV show: A crime occurs and police are called. The investigators spend a few minutes searching the room with what appears to be paintbrushes and tins of dust. Finally one of them holds up a clear plastic bag, declaring, "We got a print!"

A few minutes after that, they put the fingerprint into a computer and voilà, there's an image of the bad guy. They track him down, make an arrest, everyone high-fives each other, and the credits roll. If you're a fan of crime shows, you've probably seen these scenes or something like them a *billion* times.

So what's so special about a fingerprint, you wonder? How did a simple print left behind by the criminal lead to his eventual capture?

As you dip your fingertips into an ink pad and press them to paper, you and your lab partner will be able to examine your particular fingerprints and see what makes them special. Then you can take it to the next level and have some fun with your prints, drawing with them and seeing what you can create. This experiment is one part a study of your own fingerprints and the science behind each person's unique traits, and also one part supercool rainy-day craft idea.

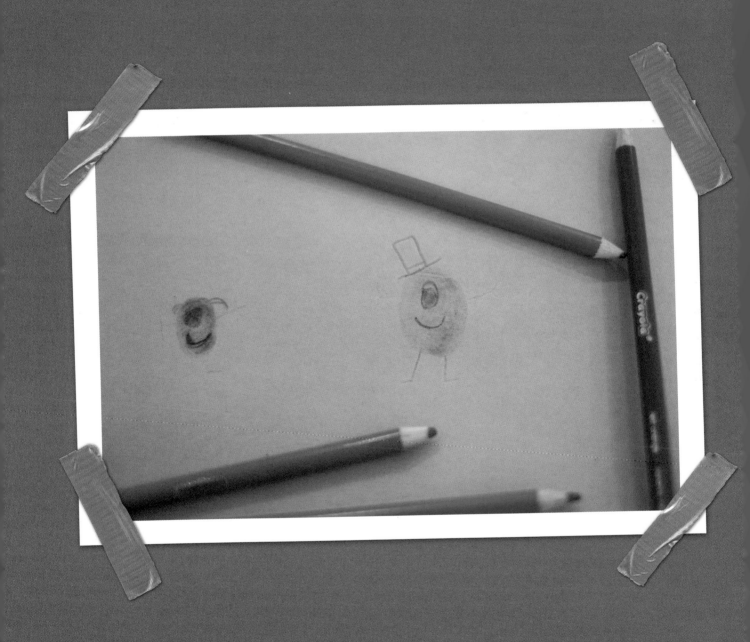

HERE'S WHY IT WORKS

Humans, monkeys, and apes all have, on the tips of our fingers, what's called friction ridge skin—our fingerprints. They're designed to help us hold onto things better—to give us a better grip. But they've also become quite handy in helping law enforcement identify the bad guys, because no two fingerprints are alike—not even those of identical twins. So if a bad guy touches something at a crime scene, police officers can "lift" those prints and try to identify who was there.

Take a close look at your fingers. Fingerprints have three main styles: loops, arches, and whorls. An arch looks like, well, an arch—like a miniature bridge almost. Loops appear to form a really tight turn in a road—a hairpin curve. Whorls form miniature spiral circles.

Now, take an even closer look. There's probably a lot more than just loops, arches, and whorls. Do you see all those tiny lines, breaks, splits, and half-formed shapes? Investigators call those *minutiae*. While you might share main pattern similarities with your parents—say, a lot of whorls, or a lot of arches—it's the addition of these tiny features, or minutiae, that make your prints unique. Add up the main pattern and all these tiny features and investigators will be able to tell the difference between you or your parents or siblings. Can you do the same?

 DID YOU KNOW?

Your prints are formed when you're still inside the uterus, usually between 10 and 15 weeks old. First your miniature hands grow what are called volar pads—or raised pads on your palms and fingertips. Eventually those pads stop growing while your hand continues to grow and form. As your hand grows, it absorbs the volar pads and basically squishes them into your palms and fingertips, creating all those ridges and furrows that will eventually make up your print patterns.

⟹ HERE'S WHAT YOU NEED

- ❑ Ink pad

- ❑ Paper

- ❑ Coloring implements

- ❑ Clear tape

- ❑ Very important: your thumb! Or any finger, really. Try not to misplace them. You will need them for this. Although if you can't seem to find them, your toes will work just fine, too. (They're unique to each person as well, but if you misplace those, I can't help you.)

HERE'S WHAT YOU DO

1 First, press your finger or thumb into the ink pad. Hopefully you have one in the art bin. (If not, lightly paint your finger pads with a marker or a light coat of house paint. Let's not let messes get in the way of cool science. Use what you have and what works.)

2 Press your inked digit to your paper. *(See Fig. 1.)*

3 Take note of the print you leave behind, because here is where you can take this science experiment/craft in a billion different directions. I like to draw faces and characters out of the miniature ridges and furrows left behind in the prints. One print might make for a great monster, while another might make the perfect family pet. You can make personalized art or thank you cards with this simple project. Or, you can just begin to look for what makes you . . . *you.* *(See Fig. 2.)*

4 Next, try "lifting" your own prints as well. After you use the ink pad and paper to make a print, press a piece of clear tape, sticky side down, onto your print. Peel the tape off and then press it down on another piece of paper. See how the print is lifted onto the tape and then carried over to the new paper? Pretty cool, right?

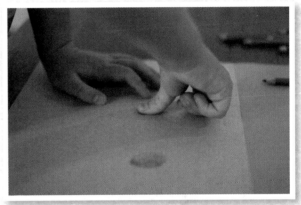

Fig. 1

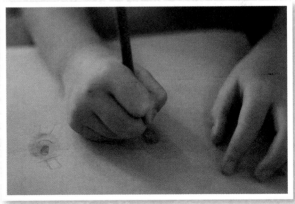

Fig. 2

STANDARD U.S./METRIC
MEASUREMENT CONVERSIONS

VOLUME CONVERSIONS	
U.S. Volume Measure	**Metric Equivalent**
⅛ teaspoon	0.5 milliliter
¼ teaspoon	1 milliliter
½ teaspoon	2 milliliters
1 teaspoon	5 milliliters
½ tablespoon	7 milliliters
1 tablespoon (3 teaspoons)	15 milliliters
2 tablespoons (1 fluid ounce)	30 milliliters
¼ cup (4 tablespoons)	60 milliliters
⅓ cup	90 milliliters
½ cup (4 fluid ounces)	125 milliliters
⅔ cup	160 milliliters
¾ cup (6 fluid ounces)	180 milliliters
1 cup (16 tablespoons)	250 milliliters
1 pint (2 cups)	500 milliliters
1 quart (4 cups)	1 liter (about)
1 gallon	3.8 liters

WEIGHT CONVERSIONS	
U.S. Weight Measure	**Metric Equivalent**
½ ounce	15 grams
1 ounce	30 grams
2 ounces	60 grams
3 ounces	85 grams
¼ pound (4 ounces)	115 grams
½ pound (8 ounces)	225 grams
¾ pound (12 ounces)	340 grams
1 pound (16 ounces)	454 grams

TEMPERATURE CONVERSIONS

Degrees Fahrenheit	Degrees Celsius
120 degrees F	49 degrees C
140 degrees F	60 degrees C
200 degrees F	95 degrees C
250 degrees F	120 degrees C
275 degrees F	135 degrees C
300 degrees F	150 degrees C
325 degrees F	160 degrees C
350 degrees F	180 degrees C
375 degrees F	190 degrees C
400 degrees F	205 degrees C
425 degrees F	220 degrees C
450 degrees F	230 degrees C

LENGTH

U.S.	Metric
1 inch	2.54 cm
1 foot	30.48 cm

INDEX

Note: Page numbers in *italics* indicate experiments.

A

Acid Rain, *139–42*
Action and reaction, law of, 90
Afterimages, 167, 168, 169
Air, taking up space, *149–53*
Air pressure, *149–53*
    about, 95, 123, 124
    blowing to create, 124
    Egg in a Bottle experiment and, *123–26*
    Floating Water experiment and, *93–97*
    speed of air and, 95
Animal Camouflage, *77–80*
Aortic arches, 61

B

Balancing
    Balancing Act, *127–31*
    Blind Balance, *161–65*
    ears and, 163
    eyes and, *161–65*
Balloons
    Balloon Toss, *103–6*
    Banana Balloon, *81–85*
    Fast Corners and, *99–101*
    helium, air pressure and, 90, 95, *99–101*
    The Space of Air and, *149–53*
    static electricity and, 115
    Straw Balloon Rocket Blasters, *89–92*
Banana Balloon, *81–85*
Batteries, magnetic fields and, *113–17*
Biology experiments, 54–85. *See also* Human body
  experiments
    about: overview of, 55
    Animal Camouflage, *77–80*
    Banana Balloon, *81–85*
    Colored Leaves (and flowers, etc.), *57–60*
    Falling Leaves, *67–70*
    Hole-y Walls, *71–75*
    Light Fright, *61–65*
Birdcage, *167–69*
Blind Balance, *161–65*
Blood
    feeling movement of, 171
    flow of, heart rate, pulse and, *171–74*
    skin and, 158, 160
Body. *See* Human body experiments
Boiling Ice, *37–39*
Book overview, 9–11, 19

C

Cage, bird in, experiment, *167–69*
Camouflage, animal, *77–80*
Capillary action (capillarity), 58, 60, 71
Car, experiment in, *99–101*
Carbon dioxide, 61, 68, 109, 136. *See also* Acid Rain
Celsius/Fahrenheit conversion chart, 181
Charles's Law, 24
Chemistry experiments, 20–53
    about: overview of, 21
    Boiling Ice, *37–39*
    Floating Grape, *27–30*
    Penny Shiners, *31–35*
    Rainbow Water Stacks/Cool Beach Lava Lamp/

Sweet, Delicious Density Treat, *47–53*

Rock Candy Crystals, *41–45*

Soap Clouds, *23–25*

Chlorophyll, 68

Coke rocket, *107–12*

Colors

Colored Leaves (and flowers, etc.), *57–60*

Colors of Light, *119–21*

Falling Leaves (why leaves change color), *67–68*

primary, 119–20, 121

of the sky, 149, 153

Conversion charts, U.S./metric, 180–81

Cool Beach Lava Lamp, *50, 52*

Copper oxide, removing, *31–35*

Crypsis, 78

Crystals, rock candy, *41–45*

Curiosity, 13–14, 15

D

Decomposition, gases from, *81–85*

Density

defined, 28

Floating Grape and, *27–30*

Rainbow Water Stacks/Cool Beach Lava Lamp/ Sweet, Delicious Density Treat, *47–53*

of solids vs. liquids vs. gases, 39

Dermis, inner, 160. *See also* Skin

E

Ears, balance and, 163

Earth science experiments, 132–53

about: overview of, 133

Acid Rain, *139–42*

Land Warmer, *143–47*

The Space of Air, *149–53*

Volcano Time!, *135–38*

Earthworms. *See* Worms

Eggs

Egg in a Bottle, *123–26*

Impulse Momentum Theorem experiment, *104*

Electricity

atoms, subatomic particles and, 115

current of, 115

how it works, 115

Magnetic Fields and, *113–17*

solenoids and, 115, 117

Electrons, 115

Energy. *See also* Electricity

conserving, 142

Magnetic Fields and, *113–17*

soil heating up (Land Warmer) and, *143–47*

Epidermis, 160. *See also* Skin

Erector pili, 158

Eruptions, volcano, *135–38*

Ethylene gas, 83, 84

Expansion

of gases, 39, 83, 84, 136, 150

Soap Clouds demonstrating, *23–25*

Experiments

biology. *See* Biology experiments; Human body experiments

chemistry. *See* Chemistry experiments

curiosity and, 13–14, 15

importance of failure, 15–16

laying groundwork for future, 17

letting kids do the work, 11

materials required for, 10–11

modifying, 10

overview of, 9–10

physics. *See* Physics experiments

planet Earth. *See* Earth science experiments

repeating to get right, 16

safety precaution, 10–11

Scientific Method and, 18–19

time requirements, 16–17

Eyes

    afterimages seen by, 167, 168, 169

    balance and, *161–65*

    bird-in-cage experiment, *167–69*

    retina, rods and cones, 168, 169

F

Fahrenheit/Celsius conversion chart, 181

Failure, importance of, 15–16

Falling Leaves, *67–70*

Fast Corners, *99–101*

Fingerprint Monsters, *175–79*

Fingerprints, about, 175–77

Floating things. *See also* Rocket propulsion

    about: helium balloons, 90, 95, 99–101

    Floating Grape, *27–30*

    Floating Water, *93–97*

    Rainbow Water Stacks/Cool Beach Lava Lamp/

        Sweet, Delicious Density Treat, *47–53*

Flowers, coloring, *57–60*

Future, these experiments and, 17

G

Gases. *See also specific gases*

    Acid Rain and, *139–42*

    from decomposition, *81–85*

    defined, 39

    expansion of, 39, 83, 84, 136, 138, 150

    in liquid (Mentos and Coke Rocket), *107–12*

    The Space of Air, *149–53*

    states of matter and, 21, 38, 39

    Volcano Time! and, *135–38*

    water becoming, 37

Goose bumps, 158

Grams and ounces, 180

Grapes, floating, *27–30*

Guidelines

    getting experiments right, 16

    importance of failure, 15–16

    letting kids do the work, 11

    materials required, 10–11, 14

    safety precaution, 11

    Scientific Method, 18–19

    time requirements, 16–17

H

Helium balloons. *See* Balloons

Hole-y Walls, *71–75*

Human body experiments, 154–79

    about: overview of, 155

    Birdcage, *167–69*

    Blind Balance, *161–65*

    Fingerprint Monsters, *175–79*

    Hot and Cold, *157–60*

    Marshmallow Pulse Keeper, *171–74*

Hypothesis, 18

I

Ice, boiling, *37–39*

Impulse Momentum Theorem (Balloon Toss),

    103–6

Inertia, Fast Corners experiment, *99–101*
Inner dermis, 160. *See also* Skin
Iodine, experiment using, *71–75*

K

Kids
    curiosity of, 13–14, 15
    failing, importance of, 15–16
    letting them do the work, 11
    safety precaution, 10–11
    STEM movement and, 17

L

Land Warmer, *143–47*
Lava lamp, *50, 52*
Laws of Motion, 89, 90, 100, 101
Leaves
    Colored Leaves (and flowers, etc.), *57–60*
    Falling Leaves, *67–70*
    why they change color, 67–68
Length conversion chart, 181
Light, colors of, *119–21*
Light Fright, *61–65*
Liquids
    Acid Rain and, *139–42*
    boiling experiment, *37–39*
    defined, 39
    density experiments with, *27–30, 47–53*
    gas in. *See* Mentos and Coke Rocket
    nitrogen, 13
    osmosis of, *71–75*
    states of matter and, 21, 38, 39
  Liters and milliliters, 180

M

Magnetic Fields, *113–17*
Marshmallow Pulse Keeper, *171–74*
Marshmallows, expansion of, *25*
Materials required, 10–11, 14. *See also specific experiments*
Matter, states of. *See also* Gases; Liquids; Plasmas; Solids
    about: overview of, 21
    boiling liquid and, 38
    defined, 39
Mentos and Coke Rocket, *107–12*
Metric/U.S. measurement conversion charts, 180–81
Milliliters, 180
Mimicry, *78*
Molecules
    acid rain and, 141
    of air, air pressure and, 124
    nucleation of, 42, 109
    seeping through holes, *71–75*
    states of. *See* Matter, states of
Momentum experiment, *103–6*
Motion, laws of, 89, 90, 100, 101
Motion dazzle, *78*
Motion sickness, 163, 165

N

Neutrons, 115
Newton, Sir Isaac, 90, 100, 101
Newton's Laws of Motion, 89, 90, 100, 101
Nitrogen, liquid, 13
Nitrogen dioxide, 149
Nucleation, 42, 109

O

Oersted, Hans Christian, 115
Organ, largest, 157
Osmosis, *71–75*
Ounces to grams, 180
Oxygen
    blood, body and, 171, 172, 174
    burning off, 124
    changing metals, 31, 33
    from plants, 61, 68

P

Pencil and ruler, balancing, *127–31*
Pennies, balancing ruler with, *127–31*
Penny Shiners, *31–35*
Perspiration, 157
Phloem, 60
Photosynthesis, 68
Physics experiments, 86–131
    about: overview of, 87
    Balancing Act, *127–31*
    Balloon Toss, *103–6*
    Colors of Light, *119–21*
    Egg in a Bottle, *123–26*
    Fast Corners, *99–101*
    Floating Water, *93–97*
    Magnetic Fields, *113–17*
    Mentos and Coke Rocket, *107–12*
    Straw Balloon Rocket Blasters, *89–92*
Pilomotor reflex, 158
Plants
    capillary action in, 58, 60, 71
    chlorophyll and, 68
    Colored Leaves (and flowers, etc.), *57–60*
    decomposition of, *81–85*
    Falling Leaves, *67–70*
    Falling Leaves (why leaves change color), *67–68*
    osmosis and, *71–75*
    phloem and xylem, 60
    photosynthesis of, 68
Plasmas, 21, 39
Protons, 115
Pulse, measuring, *171–74*
Punch recipe, 53

R

Rain, acid, *139–42*
Rainbow Water Stacks, *47–53*
Retina, rods and cones, 168, 169
Rock Candy Crystals, *41–45*
Rocket propulsion. *See also* Volcano Time!
    about: nucleation and, 109
    Mentos and Coke Rocket, *107–12*
    Straw Balloon Rocket Blasters, *89–92*
Ruler and pencil, balancing, *127–31*

S

Safety precaution, 10–11
Science
    challenging you, 16
    importance of failure and, 15–16
    STEM movement, 17
Science, Technology, Engineering, and Math (STEM), 17
Scientific Method, 18–19
Skin
    functions of, 157
    goose bumps on, 158

human, recognizing things, *157–60*

as largest organ, 157

layers of, 160

pilomotor reflex of, 158

worms breathing through, 63

Soap Clouds, *23–25*

Soil, heating up, *143–47*

Solenoids, 115, 117

Solids

defined, 39

density experiment, *27–30*

liquids and, creating gas, 136

states of matter and, 21, 38, 39

Space of Air, The, *149–53*

Straw Balloon Rocket Blasters, *89–92*

Subcutis, 160. *See also* Skin

Sulfur dioxide, 149

Supplies required, 10–11, 14. *See also specific experiments*

Sweet, Delicious Density Treat, *51, 53*

T

Temperature

Fahrenheit/Celsius conversion chart, 181

measuring, 146

skin detecting, *157–60*

of soil vs. water, *143–47*

Thermodynamics, 38, 39

Time requirements for experiments, 16–17

V

Vinegar

baking soda and, for gas expansion, *135–38*

removing copper oxide, *31–35*

salt and, to clean pennies, *31–35*

Volcano Time!, *135–38*

Volume conversion chart, 180

W

Water

absorption through plastic, *71–75*

air pressure and, 95

balloons, Balloon Toss experiment, *103–6*

Boiling Ice and, *37–39*

density of, 49. *See also* Density

Floating Water, *93–97*

hot and cold, skin detecting, *157–60*

leaves absorbing, *71–75*

osmosis of, *71–75*

Rainbow Water Stacks, *47–53*

soil heating faster than. *See* Land Warmer

Weight conversion chart, 180

Worms

blood-pumping vessels of, 61

breathing through skin, 63

castings (poo) of, 61

experiment to understand, *61–65*

length of, 63

preferences of, 63, 65

X

Xylem, 60

ABOUT THE AUTHOR

Mike Adamick is a stay-at-home dad and author of *Dad's Book of Awesome Projects*, a 2013 Amazon Book of the Year. When he's not making cool crafts with his daughter, he writes for the *San Francisco Chronicle*, KQED Radio, and Jezebel.com. His work has also appeared in the *New York Times*, *McSweeney's*, *Details* magazine, MSNBC, NPR's *Morning Edition*, the *Los Angeles Times*, and the *New York Observer*. Disney's parenting site, Babble.com, named his website "the best dad blog in cyberspace." He lives in San Francisco with his wife, Dana, and daughter, Emmeline. You can visit his website at *www .mikeadamick.com*.